Springer Series on Touch and Haptic Systems

More information about this series at http://www.springer.com/series/8786

Hasti Seifi

Personalizing Haptics

From Individuals' Sense-Making Schemas
to End-User Haptic Tools

 Springer

Hasti Seifi
University of British Columbia
Vancouver, British Columbia, Canada

ISSN 2192-2977   ISSN 2192-2985   (electronic)
Springer Series on Touch and Haptic Systems
ISBN 978-3-030-11378-0   ISBN 978-3-030-11379-7   (eBook)
https://doi.org/10.1007/978-3-030-11379-7

This Springer imprint is published by the registered company Springer Nature Switzerland AG
The registered company address is: Gewerbestrasse 11, 6330 Cham, Switzerland

For my family:
Mansoureh, Yahya, Mozhdeh, and Soroush
who live apart but are always close in my
heart.

Series Editors' Foreword

This is the sixteenth volume of 'Springer Series on Touch and Haptic Systems', which is published as a collaboration between **Springer** and the **EuroHaptics Society**.

Personalizing Haptics describes the phenomena in physical haptics interaction on perceptual terms. This work offers a new qualitative research tool based on user haptic experiences, which includes a large number of results on vibrotactile signals that have been derived by combining quantitative approaches and qualitative research methods. Moreover, it also covers crowdsourcing haptic data collection.

This study highlights the importance of choosing, tuning, and chaining the creation of stimuli to control haptic facets that reflect user cognition and emotion. The theoretical model and implementation of vibratory haptics are quite effective in developing customized applications.

This book originates from the thesis of Dr. Saifi Hasti who has received the EuroHaptics award for the Best Ph.D. Thesis in 2017. This award recognizes the relevance of this work, which could open a new path to research methods and provide reliable qualitative research tools that may benefit the whole haptics community.

January 2019

Manuel Ferre
Marc O. Ernst
Alan Wing

Preface

Synthetic haptic sensations will soon proliferate throughout many aspects of our lives, well beyond the simple buzz we get from our mobile devices. This view is widely held, as evidenced by the growing list of use cases and industry's increasing investment in haptics. However, taking haptics to the crowds will require haptic design practices to go beyond a one-size-fits-all approach, common in the field, to satisfy users' diverse perceptual, functional, and hedonic needs and preferences reported in the literature.

In this book, we tackle *end-user personalization* to leverage the utility and aesthetics of haptic signals for individuals. Specifically, we develop effective haptic personalization mechanisms, grounded in our synthesis of users' sense-making schemas for haptics. First, we propose a design space and three distinct mechanisms for personalization tools: *choosing*, *tuning*, and *chaining*. Then, we develop the first two mechanisms into: (1) an efficient interface for *choosing* from a large vibration library, and (2) three emotion controls for *tuning* vibrations. In developing these, we devise five haptic *facets* that capture users' cognitive schemas for haptic stimuli, and derive their semantic dimensions and between-facet linkages by collecting and analyzing users' annotations for a 120-item vibration library. Our studies verify the utility of the facets as a theoretical model for personalization tools.

In collecting users' perception, we note a lack of scalable haptic evaluation methodologies and present two methodologies for large-scale in-lab evaluation and online crowdsourcing of haptics. This book focuses on vibrotactile sensations as the most mature and accessible haptic technology but the proposed methods extend beyond vibrations and inform other categories of haptics.

This book presents my contributions to the field of haptics during my Ph.D. studies and several components of this research are results of collaboration with other individuals. I acknowledge the collaborative nature of the work by using the pronoun "we" throughout the book.

Vancouver, Canada
January 2019

Hasti Seifi

Acknowledgements

First and foremost, I would like to express my sincere gratitude to my Ph.D. supervisor, Dr. Karon E. MacLean, who provided supervision and feedback on all aspects of the work reported here. I would like to thank her for all the research and career mentorship I have received during these years. Karon, I have been lucky to work with you and I continue to be amazed by your great feedback and insights.

I would like to thank Dr. Katherine J. Kuchenbecker who accepted me as a postdoctoral researcher in her lively and diverse research group and supported me in many ways at a pivotal time in my career. Katherine, I hope to be a compassionate leader like you.

I am grateful to my Ph.D. supervisory committee, Dr. Tamara Munzner, and Dr. James T. Enns. I thank Dr. Munzner for her great information visualization course, which shaped part of this book, and for her constructive feedback on this research. I thank Dr. Enns for agreeing to be part of my Ph.D. and M.Sc. studies and for his thought-provoking questions and feedback in our meetings. My decision to pursue a Ph.D. was in part the result of working with him in my M.Sc. studies. I am also grateful to Drs. Seungmoon Choi, Machiel Van der Loos, and Ronald A. Rensink for their comments on an earlier version of the content that was published in my Ph.D. thesis.

I would like to give special thanks to my student collaborators. In particular, Dr. Oliver Schneider, who is now an assistant professor at the University of Waterloo, for being a wonderful colleague and friend and for all our productive discussions and teamwork. Also, special thanks to Kailun Zhang, Matthew Chun, Salma Kashani, Chamila Antonypillai, and Dilorom Pardaeva for all their contributions to this work. These collaborations enabled me to aim higher and achieve more than what would be possible by me, as an individual.

My special gratitude and friendship go to Dr. Mona Haraty, Dr. Saeedeh Ebrahimi Takalloo, and Mahsa Khalili for our discussions, and for the countless number of times I have benefited from their experience and advice in my studies.

I thank the faculty members and the students of the Designing for People (DFP) and SPIN research groups, for all their great feedback on my research, paper drafts, and practice talks. I have learned a lot by being a member of DFP. In particular, I thank the faculty members, Drs. Joanna McGrenere, Tamara Munzner, and Kellogg Booth, and the DFP graduate students, Kamyar Ardakani, Paul Bucci, Laura Cang, Matthew Chun, Jessica Dawson, Francisco Escalona, Anna Flagg, Dr. Brian Gleeson, Dr. Mona Haraty, Izabelle Janzen, Dr. Idin Karoui, Salma Kashani, Soheil Kianzad, Juliette Link, Dr. Syavash Nobarany, Antoine Ponsard, Yasaman Sefidgar, Dr. Oliver Schneider, Andrew Strang, Diane Tam, Dilan Ustek, and Kailun Zhang.

I am grateful for having many wonderful friends who made these years a memorable experience. In particular, I thank Dr. Saeedeh Ebrahimi Takalloo, Mahsa Khalili, Dr. Pooyan Fazli, Dr. Maryam Saberi, Marjan Alavi, and Bardia Aghabeigi for many joyful memories, and their emotional support in this journey.

I thank UBC's Four Year Doctoral Fellowship (4YF) program and the Natural Sciences and Engineering Research Council of Canada (NSERC) for providing the funding for this work. I am also grateful to Dr. Hong Tan and Dr. Colin Swindells who made the work reported in Chaps. 2 and 6 possible by sharing their high fidelity actuators with us.

Lastly, my immense love and gratitude go to my parents, Dr. Mansoureh Tadayoni and Dr. Yahya Seifi, and my siblings, Dr. Mozhdeh Seifi and Soroush Seifi, for their emotional support and for being my best mentors and friends even though we are apart. This book is for you.

Contents

Chapter 1
Introduction

Abstract With today's early state of haptic technology and of consumer exposure to its potential, personalization of haptic experiences may seem premature: few have experienced haptics beyond the binary on/off buzz delivered by a phone or watch. However, far greater possibility is waiting in the wings, with the haptics industry projected to expand dramatically in the coming years and industry practitioners seeking guidelines for how to design rich expressive sensations. In fact, a primary motivation for research in haptic personalization is that, first, broad uptake of the haptic modality is unlikely without personalization, because of major differences in how individuals perceive, prefer, and will ultimately utilize it. Secondly, supporting haptic personalization is not straightforward because so little is understood of how people cognitively interpret and remember touch sensations. Beginning to address this causality dilemma is the purpose of this book.

1.1 Motivation

1.1.1 Leveraging Haptic Utility

Haptic signals can convey rich information. Although most people's everyday exposure to haptics is limited to simple binary buzzes from their cellphones, studies show that rich sensations and high utility is possible [1–4]. Haptic signals can serve purely functional and informational purposes (e.g., facilitate time tracking [5], provide navigation information and guidance [3, 6, 7], support remote collaboration [1]) or enhance realism and aesthetic experience of entertainment media (e.g., multimodal interfaces [2, 8], games [9], and storytelling [4]).

However, the utility of haptic signals depends on their match to users' cognitive schemas. Although people can learn arbitrary meaning-mapping schemes [10, 11], signals that "make sense" are easier to learn and memorize, and have higher aesthetic appeal [4, 12]. In everyday physically and cognitively demanding scenarios

© Springer Nature Switzerland AG 2019

H. Seifi, *Personalizing Haptics*, Springer Series on Touch
and Haptic Systems, https://doi.org/10.1007/978-3-030-11379-7_1

(e.g., presentation, meeting, exercising), these characteristics either drive wide adoption of haptics or constrain their use to a niche group of people.

Unfortunately, designing intuitive haptic signals is a challenge [13, 14]. Due to hardware limitations, a large portion of the design space is not aesthetically appealing and many points in this space are perceptually similar. Further, despite ongoing research efforts, limited guidelines are available on affective and intuitive design. Designing intuitive signals remains an art, requiring extensive design experience as well as constant evaluation and refinement. Individual differences in experiencing haptics amplify the problem. Decades of research suggest that people differ on several levels from tactile acuity and receptors, to tactile information processing and memory, as well as preference and description of sensations [8, 15–18].

To have effective signals despite the above challenges, individuals must be able to improve personal salience by altering available designs aimed for an average user. While adjusting signal strength can address differences in tactile acuity, tweaking can go beyond that to adjust information density, signal-meaning assignment, and aesthetic qualities of the signals.

To achieve these, personalization tools must be simple and efficient. Difficult things seem fancy and become obsolete in the cost-benefit trade offs by users. In contrast, there are many examples of well-designed tools for self-expression finding a large audience. According to personalization literature in other domains, take-up improves with sense of control and identity, frequent usage, ease-of-use and ease-of-comprehension in personalization tools and is hindered by difficulty of personalization processes [19–23]. Color and photo editing tools are good examples where wide suites of tools, available for selection and editing (e.g., color swatches and gamut, preset photo filters and sliders), have led to large adoption by end-users.

In haptics, however, a large knowledge and motivation gap divides haptic professional and lay users. Existing design and authoring tools support the former group by providing control over low-level engineering parameters. For wide adoption by novice users, haptic personalization tools must be far easier to use, and this entails operating in users' perceptual and cognitive space (Fig. 1.1). We anticipate that such improvements will be valuable to haptic professionals as well: despite having the knowledge to derive haptic sensations by controlling indirect parameters, having perceptually salient "knobs" to turn will add creativity and efficiency to their process.

1.1.2 Informing Haptic Design and Evaluation

Last but not least, research on personalization can inform haptic design practices and tools. Developing simple yet effective personalization tools requires a deep understanding of common patterns in users' perception, which in turn enables effective and rich vibration design for a large audience. Further, simple and efficient authoring tools are useful for design; they enable rapid sketches and refinements, and facilitate the creative design process. The tools and guidelines we presented in this book are

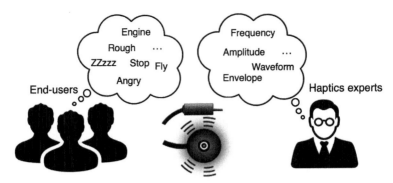

Fig. 1.1 Experts and lay users think and talk about haptic sensations differently. Experts think in terms of engineering parameters, whereas lay users describe the sensations according to their sensory and affective connotations

motivated by and contribute to both design and personalization domains. Finally, designing tools for a diverse audience requires haptic evaluation at a large scale. This requirement, in turn, highlights gaps in the tactile evaluation methodology that are not faced in typical small-scale lab-based studies. Solutions devised for those gaps expand the suite of haptic evaluation methodologies available to designers and point to future directions for research and development.

Thus, this book has three main research themes. The goal of the research presented here is to support haptic personalization (1). In doing so, we also investigate common patterns in users' perception of haptic signals (2), and devise methodologies for large-scale evaluation of haptics (3). We focus on vibrotactile stimuli, as the most mature, ubiquitous, and accessible type of haptic feedback for end-users. Technological advances and research on psycho-physical attributes, design tools, and applications for vibrotactile stimuli enable investigation of affective qualities for these sensations.

In the following, first we outline past progress in the above theme areas (Sect. 1.2), then summarize the components of this book by chapter (Sect. 1.3). Finally, we present our high-level contributions to each of the themes (thematic view in Sect. 1.4) and highlight the links between the chapters and contributions in Table 1.1.

1.2 Situating Our Work

Here, we present a brief overview of the related literature on the three main themes of this book. Focused related work sections in the following chapters will build upon this first pass, each of them emphasizing literature pertinent to their research question(s).

1.2.1 Supporting Personalization

There has been substantial personalization research in other modality and application domains, providing insights on effective mechanisms, in contrast to minimal efforts to date for haptic experience personalization.

Personalization mechanisms in other domains—Henderson and Kyng described three approaches for changing the behavior of a software tool: (1) choosing between pre-defined behaviors, (2) constructing new behaviors from existing pieces, and (3) altering an artifact through modifying the source code [24]. These approaches vary in the background knowledge and time investment required of users. In the first approach, a settings panel allows users to choose between existing configurations and add/remove toolboxes and features from the interface [25]. In the second approach, the interface provides users with a set of building blocks that they can combine for new behaviors [26, 27]. The last approach requires end-user development and programming and is typically facilitated by visual programming languages or light-weight scripting [28].

Existing commercial interfaces deploy and expand upon the above mechanisms. A suite of tools exists for choosing and adjusting colors including pre-designed palettes, color picker, and color gamut for choosing from a set as well as sliders to change RGB, brightness, hue, etc. In the photo editing domain, one can make detailed modifications (e.g., crop, select, move or rotate a region, adjust color for an individual or groups of pixels) or apply overall effects to a picture. Instagram [29], Adobe Lightroom [30], and Adobe Photoshop [31] include a suite of tools and sliders for these manipulations. Similarly, in games and virtual worlds, users can modify features of a single character (e.g., appearance, power, etc.), or configure components of an environment by choosing from a set(s) of alternatives, or adjusting sliders [32, 33]. These instances highlight the prevalence of pre-designed collections and simple tuning mechanisms for personalization in other domains.

Haptic personalization—In comparison, there exists very little support for personalization in haptics. iOS 5.0 and later versions offer users a short list of (less than 10) vibration patterns to choose from. In addition, users can create a custom vibration by tapping a pattern on the interface [34]. Besides these, two haptic collections were introduced in the last few years, offering a wide range of pre-designed sensations, each with a unique interface and organization schema.

Pre-designed haptic collections and their structure—In March 2011, Immersion Inc., a multinational company specializing in haptic technology, released a library composed of 120 vibrotactile effects and an API for accessing them [35]. Two Android applications showcase Immersion's vibration library and API to users. The first application, released in 2011, provides a list view of the effects, grouped based on their functionality or signal content (e.g., vibrations with "two clicks" are grouped together.) [35]. "Haptic Muse" was the second application, temporarily released in 2013, to showcase usage examples of the vibrations in the context of simple multimodal game scenes [36]. In 2014, Disney Research introduced their FeelEffects library which is composed of 54 sensations, grouped into six families of metaphors (e.g., rain, explosion) [4]. Vibrations in each family can be accessed through a set of presets (e.g., heavy rain, downpour, sprinkle) and sliders (e.g., drop strength, size, frequency). FeelMessenger is an instant messaging application prototype, based on the FeelEffects library, that allows users to accompany their text messages with customized vibration sensations [37].

To fill the large personalization gap in haptics, a first step is to develop effective mechanisms and tools for haptic personalization which can in turn enable future research in the area.

Adaptive approaches—A closely related topic is research on adaptive interfaces which can automatically adjust their functionality and/or content or provide recommendations based on users' preferences, interaction history, or state (e.g., location, activity, etc.) [38–40]. While adaptive interfaces eliminate the personalization effort for users, research suggests that they prefer easy-to-use personalizable systems and perceive to have higher performance with them [38]. Further, improper automatic adaptation can, in fact, lower users' performance and increase their cognitive load compared to using a static one-size-fits-all interface [41, 42]. In haptics, limited understanding of users' preferences and suitable adaption targets for different individuals makes proper adaption particularly challenging. Thus, haptic personalization research takes precedence over adaptive approaches. Our work informs future efforts on adaptive haptic systems by characterizing users' cognitive and affective schemas for haptics (Chaps. 4 and 5).

1.2.2 Understanding Common Patterns and Individual Differences

The haptics community has established foundations of haptic design. Past studies have outlined psychophysical properties of vibrations (e.g., just-noticeable difference and detection thresholds for different body locations) [43–47], identified a set of design parameters (e.g., rhythm, energy, envelope) [48–52], and provided guidelines for designing a set of perceptually distinct vibration sensations [49, 53]. However, few guidelines exist on translating high-level design descriptions (e.g., intended emotions, metaphors, or usage examples) to sensory or engineering parameters available in the authoring tools. Here, we outline efforts on devising affective guidelines and categorize various instances of individual differences reported in the literature.

Devising guidelines for affective design—Previous studies in this area have simplified the question to characterizing the link between the engineering parameters of vibrations (e.g., frequency) and the two emotion attributes of pleasantness and arousal [13, 54–56]. Vibrations with longer duration, higher energy, roughness, or envelope frequency are perceived less pleasant and more urgent [13, 56]. Sine waveform is perceived smoother than square waveform and ramped signals feel pleasant [13, 57]. Little or no guidelines exist on designing for other emotion or qualitative attributes. Further, little is known about users' cognitive schemas for vibrations, the range of qualitative and affective attributes perceived for the signals, and their underlying semantic structures.

Characterizing users' language—User descriptions of haptic sensations provide a window to the signals' affective attributes. Recent studies in this domain suggest that people use a mixed language for describing haptic sensations [13, 58–60]. Sensory and emotion attributes are used most often; Guest et al. collected a dictionary of sensory and emotion words for tactile sensations and proposed comfort and arousal as the underlying dimensions for the tactile emotion words. For tactile sensation words, the results of the MDS analysis suggested rough/smooth, cold/warm, and wet/dry orthogonal dimensions [58]. Others reported using metaphors (e.g., boat, car), usage examples (e.g., warning, stop), engineering attributes (e.g., high frequency), or vocalizations (e.g., beooo, dadada, Zzzz) for describing vibrations [13, 59–61].

We developed these into haptic *facets* (categories of attributes related to one aspect of an item), that can encapsulate users' sense-making schemas for vibrations (Chaps. 4 and 5) and thus offer an effective theoretical grounding for personalization tools.

Characterizing individual differences—Besides generalizable guidelines, designing for a large audience requires an understanding of the type(s) and extent of variations that exist around an average, aggregated perception. At least three categories of individual differences are reported in the haptic literature:

- **Sensing and perception:** Sensitivity and signal resolution of mechanoreceptors can vary among individuals leading to differences in tactile acuity, threshold, and difference detection [15, 43, 62]. These differences are more pronounced for subtle sensations such as programmable friction and can impact the perceptual space of sensations. In an old study of natural textures, Hollins et al. reported a 2D perceptual space for some participants versus a 3D space for some others [16]. Individual differences in this category are commonly investigated with psychophysical studies and avoid use of subjective components such as language terms.

- **Tactile processing and memory:** People vary in their ability to process and learn tactile stimuli [8, 17, 63, 64]. As an example, an early study on Optacon at Indiana University, a tactile reading device for blind individuals, suggested two groups of "learners" and "non-learners" in a spatio-temporal tactile matching task [17]. In a longitudinal study of tactile icon learning, participants had different learning trajectories over time [11]. Similarly, recent studies with a variable friction interface show notable differences in a set of tactile tasks [8]. Others suggest that people vary in the extent they rely on touch for information gathering or hedonic purposes [18]. Haptic processing abilities can improve with practice; visually impaired individuals develop exceptional tactile processing abilities regardless of their degree of childhood vision [65].

- **Meaning mapping and preference:** People commonly need to map abstract haptic signals to a meaning. In the absence of shared cultural connotations for haptics, mapping meaning to abstract haptic signals relies on personal experiences and sense-making schemas. Differences in describing and preference for haptic stimuli, reported in the literature, suggest individualized schemas for meaning mapping [8, 13, 66].

The last category has been studied less than the other two in the literature, contributing to the challenge of designing meaningful and aesthetic haptic icons. In this book, we contribute to the last category by reporting on the variations observed for the above meaning-mapping facets.

1.2.3 Evaluating at Scale

Developing generalizable themes and design guidelines is hard, if not impossible with small scale studies. In contrast, much more can be learnt by collecting data on a wide range of sensations from a large and heterogeneous group of users. Despite ongoing progress in haptic evaluation methodologies and metrics, there is little literature on supporting tactile evaluation at scale. Past researchers have adopted or revised existing methodology in the haptic and other domains to fill this gap. Here, we focus on studies of large sets and large participant pools.

Collecting data for a large set—Studies of large sets (>40 items) are rare in the haptic literature, partially due to lack of an effective data collection methodology. When studying large sets, feedback is commonly limited to a few ratings and/or items are divided to smaller subsets, evaluated in different sessions [52, 67, 68]. In particular, Ternes et al. devised a methodology for collecting extensive multidimensional scaling (MDS) data for a large set (84 items), and established a mathematically sound procedure for merging the results together [52, 69]. We expand on these ideas in our proposed evaluation procedure.

Crowdsourcing—In other domains, user perception is collected through online platforms such as Amazon's Mechanical Turk (MTurk) [70]. Initial studies in these domains established validity of the data collected and best practices with MTurk [71–73], enabling a wide range of studies to collect data in a fraction of time and cost compared to lab-based studies [74–76]. Haptic studies, however, are left out due to the need for specialized hardware, not available to "crowds". To utilize the MTurk platform, we need a workaround for existing hardware limitations as well as studies validating data collected with remote platforms.

In this book, we propose efficient methodologies for collecting data for a large haptic set in both lab-based (Chap. 5) and remote settings (Chap. 6).

1.3 Outline of the Chapters

This book is organized in a chronological order reflecting the steps we took in tackling haptic personalization and as such each chapter motivates and contributes to the next one. Later in this chapter (Sect. 1.4), we list the book's contributions and link them to the work reported in individual chapters in Table 1.1.

Fig. 1.2 Conceptual sketch of individual differences in affective perception of vibrations

1.3.1   Chapter 2—Linking Emotion Attributes to Engineering Parameters and Individual Differences

The first step of this research was motivated by our interest in affective design and further confirmation of the gap by the literature and industry. In a review of the haptic literature, we noted few studies on affective attributes of synthetic haptic stimuli and several reports of individual differences in haptic perception and affect. At the same time, Vivitouch (a subsidiary of Artificial Muscles Inc.) contacted our lab with an interest in designing aesthetically pleasing vibrations. Together, these shaped our first research question: *What parameters contribute to affective perception of vibrations?*

To address this, we investigated the impact of vibrations' engineering properties (specifically rhythm and frequency) on affective perception of the signals (Fig. 1.2). Further, we tested if individuals' characteristics (e.g., demographics, tactile perfor- mance) can account for differences in their perception. Results from our lab-based study showed a significant impact of engineering parameters on ratings of energy, roughness, rhythm, urgency, and pleasantness but no link to individuals' characteris- tics. Further, we noted that individual differences in haptics are nuanced and cannot be easily modelled or prescribed for in design.

1.3.2   Chapter 3—Characterizing Personalization Mechanisms

To support affective design given individual differences, we proposed a pragmatic approach: enabling people, untrained in haptics, to personalize their everyday haptic signals (e.g., notifications) for their taste and utilitarian needs. Thus, we asked: *What characteristics will make a vibration personalization tool usable?*

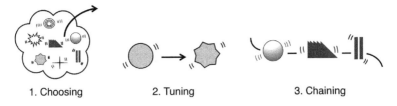

Fig. 1.3 Conceptual sketch of three personalization mechanisms for haptic sensations

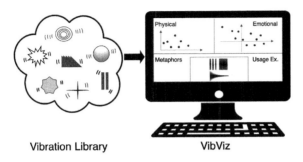

Fig. 1.4 Conceptual sketch of the *choosing* approach with VibViz

Based on a review of existing tools in haptics and other domains, we proposed five design parameters for haptic personalization tools and varied these parameters within low-fidelity prototypes of three mechanisms: (a) *choosing*: users can select from a list of pre-designed vibrations, (b) *tuning*: users can adjust high-level characteristics of a vibration by changing the value of a control, and (c) *chaining*: users combine short pre-designed tactile building blocks (e.g., by sequencing them) to create a new vibration sensation (Fig. 1.3).

Results from a Wizard of Oz (WoZ) study with paper prototypes of the tools suggested *tuning* to be the most preferred approach for being "fast", "effective", and providing a sense of "control". *Chaining* was "fun" but it required "time" and "a good mood", thus it was less practical for everyday scenarios. Finally, *choosing* was the least preferred for its limited "control" but was rated as the easiest to use. Based on the results from this study, we focused on further developing *choosing* and *tuning* as the most practical mechanisms for personalization tools.

1.3.3 Chapter 4—Choosing from a Large Library Using Facets

We conjectured that the low preference ratings for the *choosing* approach was due to the limited set of vibration options. i.e., limited control and choice. Thus, we focused

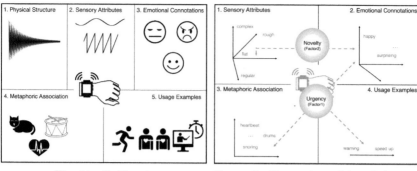

Fig. 1.5 Conceptual sketch of the five vibration facets and their underlying semantic dimensions and linkages

on providing a wide range of vibration sensations to satisfying various tastes and needs, and facilitating simple and efficient access to the library.

People unconsciously use a multiplicity of cognitive schemas to make sense of and describe qualitative and aesthetic attributes of vibrations [13, 59]. Facets and faceted browsing, from the information retrieval and library sciences literature, can encapsulate these multiple schemas. A facet includes all properties or labels related to one aspect of or perspective on an item and offers a categorization mechanism.

We compiled five haptic facets[1] based on the literature and the expertise in our research group: (1) *physical* attributes of vibrations that can be objectively measured such as duration, rhythm structure, etc. (2) *sensory* properties such as roughness, (3) *emotional* connotations, (4) *metaphors* that relate the vibration's feel to familiar examples, and (5) *usage examples* or events where a vibration fits (e.g., speed up). In parallel, we designed a library of 120 vibrations with a wide range of characteristics, and developed *VibViz*, an interactive visualization interface, that provides multiple pathways to navigate the library through the above facets (Fig. 1.4).

Results from a lab-based study confirmed utility of VibViz for searching and exploring our library. The majority of participants used and preferred the *emotion* view/facet the most but we found an interesting variation, with some preferring the other facets (e.g., usage example), and several asking for access to multiple facets.

1.3.4   Chapter 5—Deriving Semantics and Interlinkages of Facets

Confirming the facets' utility for end-users, we further investigated haptic facets to go beyond a flat list of attributes and understand their underlying semantic structures as well as the linkages between different facets (Fig. 1.5).

---

[1]Called "taxonomies" in our original conference publication.

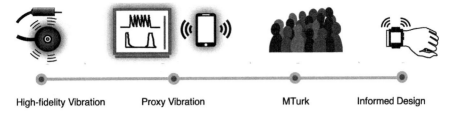

High-fidelity Vibration   Proxy Vibration   MTurk   Informed Design

Fig. 1.6 Conceptual sketch of crowdsourcing data collection for high fidelity vibrations

First, we collected annotations (ratings and tags) for the 120 vibrations in a two-stage methodology, where data from both haptics experts and lay users were combined into a final validated dataset. Next, we analyzed the annotations for their underlying semantic structure(s) and interlinkages. Specifically, we applied Multidimensional Scaling (MDS) analysis to our validated dataset, resulting in 4 *sensory*, 3 *emotion*, 2 *metaphor*, and 1 *usage example* dimension(s). Further, we investigated the linkages between the dimensions in different facets using factor analysis as well as linkages between the tags based on their co-occurrence rate in our dataset. We also reported variations, representing individual differences, in the ratings and tags for the four facets. Finally, we discussed how these results can inform three common scenarios in design and personalization of affective haptic sensations. Our dataset, source vibrations, and proposed facet dimensions were publicly released for future investigations.

1.3.5 Chapter 6—Crowdsourcing Haptic Data Collection

Our two-stage data collection methodology allowed us to collect rich information for a large library. However, it still required considerable time and effort, as well as access to haptics experts. We could collect data from a large and diverse group of users at a fraction of time and cost if we had access to crowdsourcing platforms such as Amazon Mechanical Turk. Unfortunately, haptic studies rely on specialized hardware, thus cannot be crowdsourced.

In this project, we investigated the feasibility of crowdsourcing haptic data collection using vibration proxies (Fig. 1.6). A proxy is a sensation that communicates key characteristics of a source vibration within a bounded error. We asked: *Can proxy modalities effectively communicate both engineering properties* (e.g., *duration), and high-level affective properties (roughness, pleasantness)? Can they be deployed remotely?*

To address these questions, we developed two visual proxies and a low-fidelity vibration proxy and examined them in a local lab-based as well as an online MTurk study. Results suggested that proxies are a viable approach for crowdsourcing haptics and highlighted promising directions and challenges for future work.

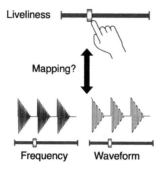

Fig. 1.7 Conceptual sketch of an emotion tuning control and its mapping to engineering attributes of vibrations

1.3.6 Chapter 7—Tuning Vibrations with Emotion Controls

Among our three personalization mechanisms, users preferred the *tuning* mechanism the most for its "ease of use" and "sense of control" (Chap. 3). Thus, in this chapter, we investigated the feasibility of designing emotion controls that allow tuning (i.e., moving) vibrations in a facet space (Fig. 1.7). We chose *agitation*, *liveliness*, and *strangeness*, the three underlying dimensions for the emotion facet (Chap. 5), as our target for emotion controls and asked: *Can we find a continuous mapping between a vibration's specific emotion property (e.g., liveliness) and its engineering parameters that apply to a diverse set of vibration patterns?*

Results from two user studies, where participants rated vibration alternatives relative to the corresponding unaltered base vibrations, suggested existence of a mapping between emotion and engineering attributes for a wide range of base vibrations. We show, based on these results, that emotion controls are automatable and discuss three example interface enabled by these.

1.4 Contributions

We started by looking at individual differences and factors that contribute to affective perception of vibrotactile stimuli, and that led us to the central goal of this book: enabling personalization of haptic sensations for end-users. We investigated haptic facets as a theoretical grounding for effective personalization tools and further developed *choosing* and *tuning* personalization tool approaches. Through our studies, we faced challenges and shortcomings in the tactile evaluation methodology and devised mechanisms to overcome those.

Our work has four major contributions: The first three pertain to the themes of supporting personalization, understanding common themes and individual differences, and evaluating at a large scale identified in Sect. 1.1. The last contribution com-

Table 1.1 The mapping from contributions to the chapters

	I-Personalization Mechanisms	II-Facets and Individual Differences	III-Evaluation Methodology	IV-Tools and Datasets
Chapter 2		Individual differences in emotion perception		
Chapter 3	Design space & three mechanisms: *choosing, tuning, chaining*			Demonstration of *choosing, tuning, chaining*
Chapter 4	*Choosing* with VibViz	Five vibrotactile facets		VibViz interface & source code
Chapter 5		Facet dimensions, linkages, & individual differences	Two-stage evaluation with experts & lay users	VibViz library & annotation dataset
Chapter 6			Crowdsourcing with proxies	
Chapter 7	*Tuning* with emotion controls	Emotion to engineering mapping		Three example tuning interfaces

prises public and open-source tools and datasets resulting from our work. We outline these contributions here, but elaborate on them in Chap. 8 (Conclusion). Table 1.1 illustrates the interleaved mapping between the chapters and contributions.

1.4.1 Effective Mechanisms for Haptic Personalization

We propose a design space for vibrotactile personalization mechanisms and develop the theoretical grounding and prototypes for two distinct mechanisms of *choosing* and *tuning* which we found to be most practical for personalization. Concrete outcomes of our progress are:

- A design space for personalization mechanisms outlined with five parameters (Chap. 3);
- Three distinct mechanisms in the above design space: *choosing, tuning,* and *chaining* (Chap. 3);
- Development of the *choosing* mechanism: an interactive library navigation interface (VibViz) and a first evaluation of its effectiveness (Chap. 4);

- Development of the *tuning* mechanism: a technical proof-of-concept on the feasibility of emotion controls and three example interfaces that can incorporate such controls (Chap. 7).

1.4.2 Haptic Facets Encapsulating Common Patterns and Variations in Affect

Realizing that *facets* could effectively structure users' cognitive processes for haptics, we compile five facets for vibrations, and characterize their attributes, underlying semantic dimensions, interlinkages, and individual differences. Our concrete contributions include:

- Five facets that encapsulate people's cognitive schemas for describing and making sense of haptic stimuli (Chaps. 4 and 5);
- Empirically derived semantic dimensions of four vibrotactile facets (Chap. 5)[2];
- Between-facet linkages at dimensional and individual tag levels, and discussion of their implications for vibrotactile design process and tools (Chap. 5);
- Mapping between emotion and engineering attributes of vibrations (Chaps. 2 and 7);
- Quantification and analysis of individual differences in rating and annotating vibrations (Chaps. 2 and 5);
- Preliminary findings on the effect of demographics, NeedForTouch (NFT) score, and tactile task performance on individual differences in affective ratings (Chap. 2).

1.4.3 Methodology for Evaluating Haptic Sensations at a Large Scale

We contribute to the tactile evaluation methodology for two cases: a) collecting rich feedback for a large stimuli set, and b) accessing crowds efficiently:

- A two-step methodology for annotating large sets of vibrotactile effects, and data on its validity and reliability (Chap. 5);
- A way to crowdsource tactile sensations (vibration proxies), with a technical proof-of-concept (Chap. 6).

---

[2]One facet is left out of the analysis as it pertains to engineering attributes of vibrations.

1.4.4 Tools and Datasets

Our work resulted in three open-source application packages and a public dataset that serve to demonstrate our contributions and support future research and developments in the area:

- Prototypes of the three personalization mechanisms for an Android phone (Chap. 3);
- VibViz (tool): A web-based interactive library navigation interface (Chap. 4);
- VibViz (dataset): Dataset of our 120-item vibration library including the vibrations' source files (.wav), annotations (facet attributes), and characterization according to the facet dimensions (Chap. 5).

References

1. Chan, A., MacLean, K.E., McGrenere, J.: Designing haptic icons to support collaborative turn-taking. Int. J. Hum. Comput. Stud. (IJHCS) **66**, 333–355 (2008)
2. Levesque, V., Oram, L., MacLean, K., Cockburn, A., Marchuk, N.D., Johnson, D., Colgate, J.E., Peshkin, M.A.: Enhancing physicality in touch interaction with programmable friction. In: Proceedings of the ACM SIGCHI Conference on Human Factors in Computing Systems (CHI '11), pp. 2481–2490. ACM (2011)
3. Brunet, L., Megard, C., Paneels, S., Changeon, G., Lozada, J., Daniel, M.P., Darses, F.: Invitation to the voyage: The design of tactile metaphors to fulfill occasional travelers' needs in transportation networks. In: IEEE World Haptics Conference (WHC '13), pp. 259–264 (2013). https://doi.org/10.1109/WHC.2013.6548418
4. Israr, A., Zhao, S., Schwalje, K., Klatzky, R., Lehman, J.: Feel effects: enriching storytelling with haptic feedback. ACM Trans. Appl. Percept. (TAP) **11**, 11:1–11:17 (2014)
5. Tam, D., MacLean, K.E., McGrenere, J., Kuchenbecker, K.J.: The design and field observation of a haptic notification system for timing awareness during oral presentations. In: Proceedings of the ACM SIGCHI Conference on Human Factors in Computing Systems (CHI '13), pp. 1689–1698. ACM, New York (2013). https://doi.org/10.1145/2470654.2466223
6. Ryu, J., Chun, J., Park, G., Choi, S., Han, S.H.: Vibrotactile feedback for information delivery in the vehicle. IEEE Trans. Haptics (ToH) **3**(2), 138–149 (2010). https://doi.org/10.1109/TOH.2010.1
7. Karuei, I., MacLean, K.E.: Susceptibility to periodic vibrotactile guidance of human cadence. In: IEEE Haptics Symposium (HAPTICS '14), pp. 141–146 (2014). https://doi.org/10.1109/HAPTICS.2014.6775446
8. Levesque, V., Oram, L., MacLean, K.E.: Exploring the design space of programmable friction for scrolling interactions. In: Proceedings of IEEE Haptic Symposium (HAPTICS '12), pp. 23–30 (2012)
9. Andrews, S., Mora, J., Lang, J., Lee, W.S.: Hapticast: a physically-based 3d game with haptic feedback. In: Proceedings of FuturePlay (2006)
10. Enriquez, M., MacLean, K.E.: The role of choice in longitudinal recall of meaningful tactile signals. In: Proceedings of Symposium on Haptic Interfaces for Virtual Environment and Teleoperator Systems, pp. 49–56 (2008). https://doi.org/10.1109/HAPTICS.2008.4479913
11. Swerdfeger, B.A.: A first and second longitudinal study of haptic icon learnability: The impact of rhythm and melody (2009)

12. Garzonis, S., Jones, S., Jay, T., O'Neill, E.: Auditory icon and earcon mobile service notifications: Intuitiveness, learnability, memorability and preference. In: Proceedings of the ACM SIGCHI Conference on Human Factors in Computing Systems (CHI '09), pp. 1513–1522. ACM, New York (2009). https://doi.org/10.1145/1518701.1518932

13. Schneider, O.S., MacLean, K.E.: Improvising design with a haptic instrument. In: Proceedings of IEEE Haptics Symposium (HAPTICS '14), pp. 327–332. IEEE (2014)

14. Schneider, O.S., MacLean, K.E.: Studying design process and example use with macaron, a web-based vibrotactile effect editor. In: Proceedings of IEEE Haptics Symposium (HAPTICS '16), pp. 52–58 (2016)

15. Lo, J., Johansson, R.S., et al.: Regional differences and interindividual variability in sensitivity to vibration in the glabrous skin of the human hand. Brain Res. **301**(1), 65–72 (1984)

16. Hollins, M., Bensmaïa, S., Karlof, K., Young, F.: Individual differences in perceptual space for tactile textures: Evidence from multidimensional scaling. Percept. Psychophys. **62**(8), 1534–1544 (2000)

17. Craig, J.C.: Vibrotactile pattern perception: extraordinary observers. Science **196**(4288), 450–452 (1977)

18. Peck, J., Childers, T.L.: Individual differences in haptic information processing: the need for touch scale. J. Consum. Res. **30**(3), 430–442 (2003)

19. Mackay, W.E.: Triggers and barriers to customizing software. In: Proceedings of ACM SIGCHI conference on Human Factors in Computing Systems (CHI '91), pp. 153–160 (1991). http://dl.acm.org/citation.cfm?id=108867

20. Marathe, S., Sundar, S.S.: What drives customization?: control or identity? In: Proceedings of ACM SIGCHI Conference on Human Factors in Computing Systems (CHI '11), pp. 781–790 (2011). http://dl.acm.org/citation.cfm?id=1979056

21. Oh, U., Findlater, L.: The challenges and potential of end-user gesture customization. In: Proceedings of ACM SIGCHI Conference on Human Factors in Computing Systems (CHI '13), pp. 1129–1138 (2013). http://dl.acm.org/citation.cfm?id=2466145

22. Nurkka, P.: "Nobody other than me knows what i Want": customizing a sports watch. In: P. Kotz, G. Marsden, G. Lindgaard, J. Wesson, M. Winckler (eds.) Proceedings of Human-Computer Interaction (INTERACT '13), vol. 8120 in Lecture Notes in Computer Science, pp. 384–402. Springer, Berlin (2013). http://link.springer.com/chapter/10.1007/978-3-642-40498-6_30

23. Blom, J.O., Monk, A.F.: Theory of personalization of appearance: Why users personalize their pcs and mobile phones. J. Hum.-Comput. Interact. **18**(3), 193–228 (2003). http://www.tandfonline.com/doi/abs/10.1207/S15327051HCI1803_1

24. Henderson, A., Kyng, M.: There's no place like home: Continuing design in use. In: Design at Work: Cooperative Design of Computer Systems, pp. 219–240. Lawrence Erlbaum Associates Inc. (1992)

25. Ponsard, A., McGrenere, J.: Anchored customization: Anchoring settings to the application interface to afford customization. In: Proceedings of the ACM SIGCHI Conference on Human Factors in Computing Systems (CHI '16), pp. 4154–4165. ACM, New York (2016). https://doi.org/10.1145/2858036.2858129

26. Fischer, G., Scharff, E.: Meta-design: design for designers. In: Proceedings of the ACM Conference on Designing Interactive Systems (DIS '00), pp. 396–405. ACM, New York (2000). https://doi.org/10.1145/347642.347798

27. Haraty, M., McGrenere, J.: Designing for advanced personalization in personal task management. In: Proceedings of the ACM Conference on Designing Interactive Systems (DIS '16), pp. 239–250. ACM, New York (2016). https://doi.org/10.1145/2901790.2901805

28. Lieberman, H., Paternò, F., Klann, M., Wulf, V.: End-user development: An emerging paradigm. In: End User Development, pp. 1–8. Springer, Berlin (2006)

29. Facebook, Inc.: Instagram. https://www.instagram.com/?hl=en. Accessed 21 Oct 2016

30. Evening, M.: The Adobe Photoshop Lightroom 5 Book: The Complete Guide for Photographers. Pearson Education, London (2013)

31. Adobe Systems, Inc.: Adobe photoshop. https://www.adobe.com/ca/products/photoshop.html. Accessed 23 Oct 2016

32. Ducheneaut, N., Wen, M.H., Yee, N., Wadley, G.: Body and mind: A study of avatar person-alization in three virtual worlds. In: Proceedings of the ACM SIGCHI Conference on Human Factors in Computing Systems (CHI '09), pp. 1151–1160. ACM, New York (2009). https://doi.org/10.1145/1518701.1518877

33. Kwak, D.H., Clavio, G.E., Eagleman, A.N., Kim, K.T.: Exploring the antecedents and consequences of personalizing sport video game experiences. Sport Mark. Quart. **19**(4), 217–225 (2010). http://ezproxy.library.ubc.ca/login?url=http://search.proquest.com/docview/851541557?accountid=14656. Copyright - Copyright Fitness Information Technology, A Division of ICPE West Virginia University Dec 2010; Document feature - Tables; Last updated - 2012-07-06

34. Wawro, A.: How to use custom vibrations in iOS 5 | PCWorld. http://www.pcworld.com/article/242238/how_to_use_custom_vibrations_in_ios_5.html. Accessed 24 Sept 2013

35. Immersion Corporation: Haptic effect preview. http://www2.immersion.com/developers/, https://play.google.com/store/apps/details?id=com.immersion.EffectPreview&hl=en. Accessed 24 Jan 2015

36. Immersion Corporation: Haptic muse. http://www2.immersion.com/developers/, http://ir.immersion.com/releasedetail.cfm?ReleaseID=776428. Accessed 24 Jan 2015

37. Israr, A., Zhao, S., Schneider, O.: Exploring embedded haptics for social networking and interactions. In: CHI '15 Extended Abstracts on Human Factors in Computing Systems (CHI EA '15), pp. 1899–1904. ACM, New York (2015). https://doi.org/10.1145/2702613.2732814

38. Findlater, L., McGrenere, J.: A comparison of static, adaptive, and adaptable menus. In: Proceedings of the SIGCHI Conference on Human Factors in Computing Systems (CHI '04), pp. 89–96. ACM, New York (2004). https://doi.org/10.1145/985692.985704

39. Gajos, K.Z., Czerwinski, M., Tan, D.S., Weld, D.S.: Exploring the design space for adaptive graphical user interfaces. In: Proceedings of the Working Conference on Advanced Visual Interfaces (AVI '06), pp. 201–208. ACM, New York (2006). https://doi.org/10.1145/1133265.1133306

40. Jameson, A.: Adaptive interfaces and agents. Hum.-Comput. Interact.: Des. Issues Solut. Appl. **105**, 105–130 (2009)

41. Mitchell, J., Shneiderman, B.: Dynamic versus static menus: an exploratory comparison. SIGCHI Bull. **20**(4), 33–37 (1989). https://doi.org/10.1145/67243.67247

42. Findlater, L., Gajos, K.Z.: Design space and evaluation challenges of adaptive graphical user interfaces. AI Mag. **30**(4), 68 (2009)

43. Stevens, J.C., Choo, K.K.: Spatial acuity of the body surface over the life span. Somatosens. Motor Res. **13**(2), 153–166 (1996)

44. Hoggan, E., Anwar, S., Brewster, S.A.: Mobile multi-actuator tactile displays. In: International Workshop on Haptic and Audio Interaction Design, pp. 22–33. Springer, Berlin (2007)

45. Pongrac, H.: Vibrotactile perception: examining the coding of vibrations and the just noticeable difference under various conditions. Multimed. Syst. **13**(4), 297–307 (2008). https://doi.org/10.1007/s00530-007-0105-x

46. Jones, L.A., Sarter, N.B.: Tactile displays: guidance for their design and application. Hum. Factors: J. Hum. Factors Ergon. Soc. **50**(1), 90–111 (2008)

47. Karuei, I., MacLean, K.E., Foley-Fisher, Z., MacKenzie, R., Koch, S., El-Zohairy, M.: Detecting vibrations across the body in mobile contexts. In: Proceedings of the SIGCHI Conference on Human Factors in Computing Systems (CHI '11), pp. 3267–3276. ACM, New York (2011)

48. van Erp, J.B., Spapé, M.M.: Distilling the underlying dimensions of tactile melodies. Proc. Eurohaptics Conf. **2003**, 111–120 (2003)

49. MacLean, K., Enriquez, M.: Perceptual design of haptic icons. In: Proceedings of EuroHaptics Conference, pp. 351–363 (2003)

50. Brown, L.M., Brewster, S.A., Purchase, H.C.: Tactile crescendos and sforzandos: Applying musical techniques to tactile icon design. In: CHI'06 Extended Abstracts on Human factors in Computing Systems (CHI EA '06), pp. 610–615. ACM (2006)

51. Hoggan, E., Brewster, S.: Designing audio and tactile crossmodal icons for mobile devices. In: Proceedings of the 9th ACM International Conference on Multimodal Interfaces (ICMI '07), pp. 162–169. ACM (2007)

52. Ternes, D.R.: Building large sets of haptic icons: Rhythm as a design parameter, and between-subjects mds for evaluation. Ph.D. thesis, The University of British Columbia (2007)
53. MacLean, K.E.: Foundations of transparency in tactile information design. IEEE Trans. Haptics (ToH) **1**(2), 84–95 (2008)
54. Koskinen, E., Kaaresoja, T., Laitinen, P.: Feel-good touch: Finding the most pleasant tactile feedback for a mobile touch screen button. In: Proceedings of the 10th International Conference on Multimodal Interfaces (ICMI '08), pp. 297–304. ACM, New York (2008). https://doi.org/10.1145/1452392.1452453
55. Zheng, Y., Morrell, J.B.: Haptic actuator design parameters that influence affect and attention. In: Proceedings of IEEE Haptics Symposium (HAPTICS '12), pp. 463–470. IEEE (2012)
56. Yoo, Y., Yoo, T., Kong, J., Choi, S.: Emotional responses of tactile icons: Effects of amplitude, frequency, duration, and envelope. In: Proceedings of IEEE World Haptics Conference (WHC'15), pp. 235–240 (2015). https://doi.org/10.1109/WHC.2015.7177719
57. O'Sullivan, C., Chang, A.: An Activity Classification for Vibrotactile Phenomena, pp. 145–156. Springer, Berlin (2006). https://doi.org/10.1007/11821731_14
58. Guest, S., Dessirier, J.M., Mehrabyan, A., McGlone, F., Essick, G., Gescheider, G., Fontana, A., Xiong, R., Ackerley, R., Blot, K.: The development and validation of sensory and emotional scales of touch perception. Atten. Percept. Psychophys. **73**(2), 531–550 (2011)
59. Obrist, M., Seah, S.A., Subramanian, S.: Talking about tactile experiences. In: Proceedings of the ACM SIGCHI Conference on Human Factors in Computing Systems (CHI '13), pp. 1659–1668. ACM (2013)
60. Doizaki, R., Watanabe, J., Sakamoto, M.: A system for evaluating tactile feelings expressed by sound symbolic words. In: Auvray, M., Duriez, C., (eds.) Haptics: Neuroscience, Devices, Modeling, and Applications: Proceedings of Eurohaptics, pp. 32–39. Springer, Berlin (2014). https://doi.org/10.1007/978-3-662-44193-0_5
61. Watanabe, J., Hayakawa, T., Matsui, S., Kano, A., Shimizu, Y., Sakamoto, M.: Visualization of tactile material relationships using sound symbolic words. In: Isokoski, P., Springare, J. (eds.) Proceedings of EuroHaptics Conference, pp. 175–180. Springer, Berlin (2012). https://doi.org/10.1007/978-3-642-31404-9_30
62. Stevens, J.C.: Aging and spatial acuity of touch. J. Gerontol. **47**(1), P35–P40 (1992)
63. Epstein, W., Hughes, B., Schneider, S.L., Bach-y Rita, P.: Perceptual learning of spatiotemporal events: Evidence from an unfamiliar modality. J. Exp. Psychol.: Hum. Percept. Perform. **15**(1), 28 (1989). http://psycnet.apa.org/journals/xhp/15/1/28/
64. Gallace, A., Spence, C.: The cognitive and neural correlates of tactile memory. Psychol. Bull. **135**(3), 380 (2009)
65. Goldreich, D., Kanics, I.M.: Tactile acuity is enhanced in blindness. J. Neurosci. **23**(8), 3439–3445 (2003)
66. Alter, A.L., Oppenheimer, D.M.: Uniting the tribes of fluency to form a metacognitive nation. Personal. Soc. Psychol. Rev. **13**(3), 219–235 (2009)
67. Spence, I., Domoney, D.W.: Single subject incomplete designs for nonmetric multidimensional scaling. Psychometrika **39**(4), 469–490 (1974). https://doi.org/10.1007/BF02291669
68. Tsogo, L., Masson, M., Bardot, A.: Multidimensional scaling methods for many-object sets: A review. Multivar. Behav. Res. **35**(3), 307–319 (2000)
69. Ternes, D., Maclean, K.E.: Designing large sets of haptic icons with rhythm. In: Haptics: Perception, Devices and Scenarios, pp. 199–208. Springer, Berlin (2008)
70. Amazon.com Inc.: Amazon Mechanical Turk Requester Best Practices Guide (2015)
71. Kittur, A., Chi, E.H., Suh, B.: Crowdsourcing user studies with mechanical turk. In: Proceedings of the SIGCHI Conference on Human Factors in Computing Systems (CHI '08), pp. 453–456. ACM, New York (2008). https://doi.org/10.1145/1357054.1357127
72. Heer, J., Bostock, M.: Crowdsourcing graphical perception: Using mechanical turk to assess visualization design. In: Proceedings of the SIGCHI Conference on Human Factors in Computing Systems (CHI '10), pp. 203–212. ACM, New York (2010). https://doi.org/10.1145/1753326.1753357

73. Mason, W., Suri, S.: Conducting behavioral research on amazon's mechanical turk. Behav. Res. Methods **44**(1), 1–23 (2012). https://doi.org/10.3758/s13428-011-0124-6
74. Chilton, L.B., Little, G., Edge, D., Weld, D.S., Landay, J.A.: Cascade: crowdsourcing taxonomy creation. In: Proceedings of the SIGCHI Conference on Human Factors in Computing Systems (CHI'13), pp. 1999–2008. ACM, New York (2013). https://doi.org/10.1145/2470654.2466265
75. Xu, A., Huang, S.W., Bailey, B.: Voyant: Generating Structured Feedback on Visual Designs Using a Crowd of Non-Experts. In: ACM Conference on Computer-Supported Cooperative Work and Social Computing (CSCW '14), pp. 1433–1444. ACM Press, New York (2014). https://doi.org/10.1145/2531602.2531604
76. Siangliulue, P., Arnold, K.C., Gajos, K.Z., Dow, S.P.: Toward Collaborative Ideation at Scale - Leveraging Ideas from Others to Generate More Creative and Diverse Ideas Pao. In: Proceedings of ACM Conference on Computer-Supported Cooperative Work and Social Computing (CSCW '15), pp. 937–945. ACM Press, New York (2015). https://doi.org/10.1145/2675133.2675239

Chapter 2
Linking Emotion Attributes to Engineering Parameters and Individual Differences

Abstract Affective response can dominate users' reactions to the synthesized tactile sensations that are proliferating in today's handheld and gaming devices, yet it is largely unmeasured, modelled or characterized. A better understanding of user perception will aid the design of tactile behavior that engages touch, with an experience that satisfies rather than intrudes. Here, we made a first attempt at developing guidelines for affective vibration design. We measured 30 subjects' affective response to vibrations varying in rhythm and frequency, then examined how differences in demographic, everyday use of touch, and tactile processing abilities contribute to variations in affective response. To this end, we developed five affective and sensory rating scales and two tactile performance tasks, and also employed a published 'Need for Touch' (NFT) questionnaire. Subjects' ratings, aggregated, showed significant correlations among the five scales and significant effect of the signal content (rhythm and frequency). Ratings varied considerably among subjects, but this variation did not coincide with demographic, NFT score, or tactile task performance. These results suggest a link between emotion and engineering parameters but highlight that individual differences in emotion perception are nuanced and cannot be modelled based on user performance or background.

2.1 Introduction

Touch is an important means of obtaining information about objects, but it is also highly connected to our emotions [1]; as a consequence, affective reactions are influential in the many small decisions we make about the objects that surround us. Only a few studies have investigated affective response to touch stimuli of any kind [2–5]; but affective study of synthetic tactile stimuli such as vibrations or variable friction is even more sparse.

While the programmable synthetic stimuli available to interaction designers are currently far less expressive than natural textures, growing attention to surface interaction in recent years means tactile technology is evolving rapidly. Already designers need to optimize its affective potential. However, we lack relevant measures and methodology for quantifying tactile affect. A multidimensional picture of subjects'

© Springer Nature Switzerland AG 2019

H. Seifi, *Personalizing Haptics*, Springer Series on Touch
and Haptic Systems, https://doi.org/10.1007/978-3-030-11379-7_2

Fig. 2.1 Individual differences in affective perception of vibrations

opinions will help reveal *patterns* of preference more effectively than can a single preference measure.

There is also a dearth of data on individualized responses. Affect studies have typically reported only responses averaged over subjects [2, 6]. There is tantalizing evidence that such variances may be substantial: e.g., Levesque et al.'s findings for subjects' preference for different patterns of variable friction [7]. Tactile designers must understand this variation's extent and driving factors.

Evidence from the literature and our own early analyses suggest that differences in everyday touch behavior, tactile abilities, and demographics might explain substantial affective response variation. A recently developed scale ('Need for Touch' (NFT)) assesses individual differences in extracting and using haptic information for everyday pleasure or utility evaluation [8]). Tactile task performance, employed as an indicator of tactile memory and processing resources, also can vary considerably across subjects [7, 9, 10]; are functional touch ability and hedonic preferences linked?

Together, these factors raise questions about the relation of demographics, NFT scores and tactile task performance to variations in affective response. Long-term, we aim to optimize and validate a set of rating scales which reflect relevant dimensions of subjective response to tactile sensations; link affective and sensory perception of tactile technology parameters (e.g., frequency, amplitude); and assess the individual differences in affect and perception and parameters that contribute to these differences.

Here, we more specifically ask: what are the relevant dimensions for measuring affective response, and can we integrate multiple rating dimensions? How does the vibration design space impact affective response? How is affective response linked to demographics, NFT scores, and tactile task performance? (Fig. 2.1) Below, we discuss these questions in light of our study results.

For maximum vibrotactile expressivity, we used a recent electroactive polymer (EAP) display from Vivitouch [11]. We examined 30 subjects' affective ratings of 1s vibrations (e.g., alerts and notifications). The rating scales, tactile stimuli and tasks were drawn from the literature and refined via pilot studies. The main study used five rating scales to examine the effect of the vibration parameters and individual differences on the subjective ratings for vibrations. The contributions of this work are:

- An initial examination of five proposed affective and sensory dimensions for rating tactile sensations (thorough validation requires further study);
- Qualitative and quantitative data on the effect of rhythm pattern and frequency on affective and sensory ratings;
- Quantitative data on individuals' variation in time and frequency matching performance;
- Preliminary findings on the effect of demographic, NFT, and tactile task performance on variations in affective ratings.

In the following we describe our apparatus, and the design and selection of the vibrations, tactile tasks and affective and sensory rating scales we used (Sect. 2.3). We report the main study and its results (Sect. 2.4), then discuss our findings and outline future work.

2.2 Related Work

2.2.1 Affective Evaluation

The touch literature lacks a consistent vocabulary for affective response. Guest et al. recently collated a large list of emotion and sensation words describing tactile stimuli [12]; then, based on Multi-Dimensional Scaling (MDS) analysis of similarity ratings, proposed *comfort* and *arousal* as underlying dimensions for the tactile emotion words, and *rough/smooth, cold/warm*, and *wet/dry* for sensation. We founded our affective rating scales on these words.

Study of affective reaction to natural stimuli [2, 3, 5] revealed dependencies on many factors, such as materials and body sites, preventing generalizations [2]. Swindells et al. obtained valence and arousal response to touching various natural materials. Comparing self report ratings and physiological recordings from subjects' bodies (EMG and skin conductance), they found self report more sensitive in discriminating the subtle affective variations to these stimuli [3]. Others have examined affective reaction to synthetic stimuli in a variety of contexts [6, 7]. Most relevantly, Takahashi et al. studied feelings of pleasantness and animacy for low frequency vibrations (0.5 to 50 Hz) applied to finger tips and wrists of six subjects [4]. They found a significant effect of frequency on animacy but no effect on pleasantness. They also found an inverted-U relation between ratings of pleasantness and animacy. Swindells et al. studied the link between the utility of various haptic feels and subjects' preference for the feel, in the context of a Fitts' law targeting task and without it. In some cases, subjects preferred the feedback providing inferior task utility [3]. In contrast, here we examine the relation of affective ratings to human tactile *abilities* rather than feedback utility.

2.2.2 Vibrotactile Stimuli

Past studies have examined the impact of several parameters on information transfer, salience, and learnability of vibrotactile icons; these include frequency, rhythm, waveform, and texture [13, 14]. These parameters are also promising candidates to evaluate in terms of their affective properties.

2.2.3 Tactile Tasks

Both sensory acuity and tactile processing resources, such as tactile working memory, contribute to a person's tactile abilities. Examination of tactile acuity for different demographics and for various body locations has shown that acuity is lower in sighted individuals and declines in old age [15]. However, acuity and Just Noticeable Difference (JND) studies did not report major individual differences [15, 16]. On the other hand, tactile individual differences were reported in some studies involving remembering or processing of tactile stimuli [7, 9, 10]. Thus, we focused here on the tasks involving tactile working memory.

Most short-term or working memory evaluation has focused on visual (iconic memory) and auditory (echoic) stimuli. A few studies have investigated time and capacity constraints of haptic working memory using tasks such as delayed matching-to-sample task or n-back task (see [17] for a review). These report 5–10 s of sensory memory, which is consistent with our observations.

2.2.4 Individual Differences in Tactile Task Performance

Considerable individual differences in tactile tasks have been reported in the literature [7, 9, 10, 18]. An early study on vibrotactile pattern recognition with the Optacon [9] found four distinct groups based on subjects' performance in three tactile tasks and their overall pattern of learning. The grouping remained consistent across the tasks and two participant pools. Another study reported two groups of learners and non-learners in a spatio-temporal pattern matching tactile task [10]. Non-learners showed little improvement over four task sessions (400 trials), while learners had better initial performance and improved. Another study with variable friction feedback showed considerable individual differences in task performance and found various preferences for different friction patterns [7]. Finally, there is evidence of individual differences in texture perception [18]. An MDS analysis on a texture similarity rating task suggested a three-dimensional space for some participants, two-dimensional for others.

In everyday life, people vary in the extent that they seek information through touch or use it for sensory pleasure [8]. 'Need for Touch' (NFT) is a 12-item questionnaire developed for consumer research that measures these differences on dimensions of pleasure (Autotelic) and information (Instrumental) touch [8]. An example Autotelic item on the questionnaire is "Touching products can be fun", whereas, "I place more trust in products that can be touched before purchase" is an Instrumental item. NFT is based on motivational differences among individuals in using touch, whereas scores on a tactile task show tactile ability differences among individuals.

Later studies have shown that higher NFT individuals have greater memory access to haptic information, seek and use it more for forming judgments [8]. These NFT studies used a relatively large number of subjects (60–100); our 30-subject exploratory trial provided less power than it required, but we included the NFT questionnaire to get an estimate of its effect size and to determine its utility for future research.

2.3 Design of Setup and Assessment Tools

In this section, we describe our apparatus and the vibrations, tactile tasks and rating scales used in our main experiment.

2.3.1 Apparatus

We used an EAP vibrotactile actuator from Vivitouch, a subsidiary of Artificial Muscles Inc. [11]. The module translates an input audio waveform to a tactile output, with an effective range of 20–200 Hz. Biggs et al. empirically modeled the actuator performance and the resulting fingerpad and palmar sensations [19]), estimating a palmar stimulation of approximately 22 dB for 75 Hz and 175 Hz, and 29 dB at 125 Hz, with a peak of 32 dB at 100 Hz. For our prototype (Fig. 2.2), we sandwiched the actuator between two thin rectangular plastic plates, each 0.5 mm × 12.5 cm × 6 cm; and encased the assembly in a protective case with same size, shape and markings of a smartphone. The prototype's total mass was 64 g.

2.3.2 Stimuli Design

Focusing on vibratory stimuli, we wanted to know which parameters could most impact subjective response and to choose a relevant range. In pilots, subjects showed some patterns of preference for longer vibrations (1s for alerts and notifications)

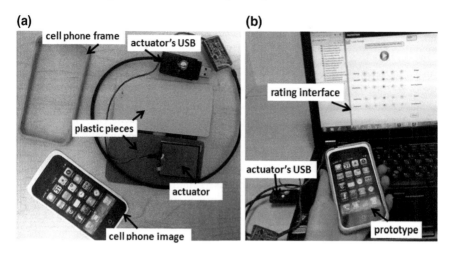

Fig. 2.2 Actuator (a), and prototype and setup (b) for the study

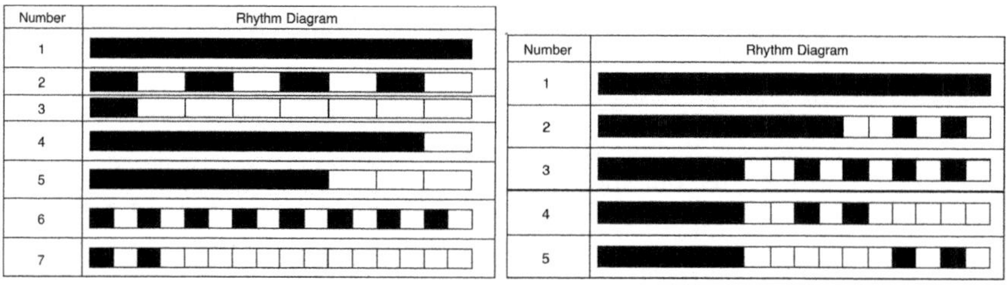

(a) Affective rating rhythm patterns   (b) Tactile task rhythm patterns

Fig. 2.3 Rhythm patterns in our user study chosen from [13]. Filled slots represent a vibration; unfilled slots represent silence or pause

compared to no preference among various short vibrations (0.1–0.3 s for keypress feedback). Thus, we focused on 1s signals. Follow-up pilots with a large set of simple and complex waveforms suggested the importance of frequency and temporal (rhythmic) pattern on subjects' preference. Base frequencies of 75 and 175 Hz captured variations in subjects' preference for different actuator frequencies in pilots; for rhythmic pattern, we drew from a perceptually validated set of rhythmic icons [13].

For our main study, we chose seven representative patterns from this rhythm set [13] (Fig. 2.3a). The patterns were each 1s, rendered in two frequencies (75 and 175 Hz), and repeated twice (7 patterns × 2 frequencies × 2 repetitions = 28 ratings per subject).

2.3.3 Tactile Task Design

We wanted to know if subjective ratings for vibrations would be affected by tactile abilities. Studies in other domains (e.g., music) have shown that proficiency with stimuli influences an individual's pattern of preference for the stimuli [20]. Also, research in processing fluency indicates a link between information processing and affective response [21]: people provided more positive affective ratings for easier-to-process stimuli, e.g., with slightly higher contrast. In addition, our post-hoc analysis of data from [7] suggested that subjects preferred friction patterns that they were better at detecting; and subjects with better performance provided twice as many positive ratings as lower-performing subjects. Clearly, tactile processing abilities *may* contribute to affective response.

For our purpose, a tactile task must predominantly detect tactile abilities (as opposed to general cognitive abilities, such as intelligence); i.e., have construct validity. It must engage tactile memory and processing resources since simple tactile acuity or JND tasks did not show considerable performance variations among subjects in past studies (Sect. 2.2). Finally, it must have a difficulty level that reveals individual differences, and be reliable enough to allow between-subject comparison. We are not aware of a standard battery of tasks that satisfies these criteria. There is one, however, for visual processing [22], and thus our task design was guided by this as well as the touch literature.

We examined rhythm, amplitude, time, and frequency matching tasks in which subjects matched a vibration to an available choice. Choices varied in rhythm, amplitude, time, or frequency. In pilot studies, rhythm matching did not rely on tactile abilities (lack of construct validity) and amplitude matching performance revealed very small individual variation. Time and frequency matching more closely met our criteria.

In our main study, tactile tasks comprised stimulus sets and a protocol. The stimulus set for both time and frequency matching tasks consisted of five rhythm patterns (Fig. 2.3b). *Time matching task (two alternative forced choice, 2AFC):* each rhythm was rendered at 75 Hz and durations of 1 and 1.3 s (pilots suggested 0.3 s difference was appropriately difficult). *Frequency matching task (3AFC):* the same five rhythms were each rendered at 75, 125 and 175 Hz and a duration of 1 s.

The same procedure was used for both tasks. For each choice we asked subjects to indicate their confidence in the answer by choosing "Maybe" (for a score of 1 or −1, for correct and incorrect matching respectively) or "Sure" (2 or −2) (Fig. 2.4) [23]. In each trial, subjects could feel the stimulus and the matching choices exactly once and were instructed to go through the choices from left to right to maintain control over order effects. Stimuli were presented in a random order and subjects were told that their choices differed in the feeling (frequency) or the timing of the vibrations.

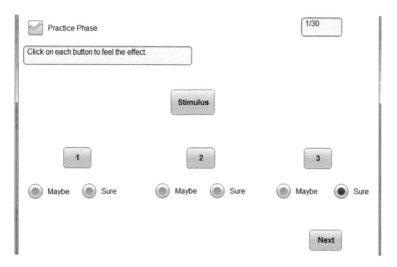

Fig. 2.4 Interface for frequency matching task (similar interface for the time matching task but with two selection buttons)

2.3.4 Affective Rating Scales Design

Most affective haptics studies have used a single measure of affective response (e.g., liking, pleasantness) or a set of self-selected scales [2, 5, 24]. An ideal affect measurement scale for our purpose must capture important dimensions of affect and perception, allow integrated analysis of those dimensions and examination of individuals' variations from average patterns of ratings, and ideally accommodate diverse tactile sensations including synthetic and natural stimuli. An integrated rating scale could also guide the design of new tactile sensations by revealing unexplored parts of the affect and sensation space based on subjects' ratings. In our discussion, we outline our progress towards these criteria, and identify future steps required for validation and further development of the scales. Nevertheless, the criteria for a desirable scale evolve as we further study affective response to tactile sensations. In the following, we use 'rating dimensions' and 'scales' interchangeably.

As a first step towards such an integrated scale, we designed an initial set of subscales based on the touch vocabulary derived by Guest et al. (see Related Work [12]). We chose a representative word from each part of their resultant emotion and sensation spaces, resulting in *unpleasant/pleasant, uncomfortable/comfortable*, and *boring/exciting* for emotion. From their sensation space, after removing words which our hardware cannot literally render (e.g., cold/warm, and wet/dry), we were left with *smooth/rough* and *soft/hard*. We added *weak/strong* and *non-rhythmic/rhythmic* to better capture the characteristics of our vibrations. This resulted in eight initial scales: *weak/strong, smooth/rough, soft/hard, non-rhythmic/rhythmic, boring/exciting, unpleasant/pleasant, uncomfortable/comfortable, dislike/like*.

Fig. 2.5 The user interface for the affective ratings

In a pilot, 6 subjects (4 males) used these scales to rate vibrations described in Sect. 2.3.2, using the interface shown in Fig. 2.5. We removed the liking and comfort dimensions because of high correlation with pleasantness ($r = 0.8$). We also removed the *soft/hard* dimension as subjects had difficulty in attributing hardness to the vibrations. Further, we re-labeled the *boring/exciting* to *calm/alarming* to achieve neutral valence and avoid inconsistent interpretations. Although not deliberate, *unpleasant/pleasant* and *calm/alarming* dimensions map to well-known valence and arousal dimensions for emotions.

This resulted in five dimensions employed in the main study: three sensory (*weak/strong, smooth/rough, non-rhythmic/rhythmic*) and two affective (*calm/alarming, unpleasant/pleasant*).

2.4 Study

2.4.1 Procedure

30 subjects participated in a one-hour, 3-part study and were compensated with $10. (1) Subjects completed a general information questionnaire and the 'Need for Touch' survey; then (2) rated 28 vibrations (Sect. 2.3.2) each on five affective and sensory scales. Vibration presentation order was randomized across subjects. On the rating interface, labels were randomly placed on the left or right side of each

scale for each subject to reduce rating bias. (3) Subjects completed two rounds of the time and frequency matching tasks (Sect. 2.3.3). Time and frequency tasks were interleaved and their order counterbalanced among subjects. Subjects held and felt the cell phone prototype in the non-dominant hand and listened to white noise to mask actuator noise.

2.4.2   Results and Analysis

Subjects were diverse. All subjects were students between 18–45 years old, 15 female, 3 left-handed, 15 from computer science and 15 from psychology, arts, chemistry etc. Sixteen participants (16) were from North America or Europe, 14 from Asia and Middle East. Fourteen participants had more than two years of musical background, six had less than two years and ten reported none. Eleven used eye glasses, and no one reported tactile deficiency. Touch tablets and smart phones, guitar, piano, Wii, and Dictaphone were mentioned as frequently used touch devices. NFT scores varied from -25 to $+30$. Following the same procedure as [8], we used a median split on NFT scores to divide the subjects into high and low NFT groups.

Rating scales revealed correlations. Overall, *smooth/rough*, *calm/alarming*, and *unpleasant/pleasant* ratings were significantly correlated. The bivariate Pearson correlation of the five ratings for all subjects showed medium significant correlation between *smooth* and *pleasant* ($r = 0.53$), *rough* and *alarming* ($r = 0.42$), *unpleasant* and *alarming* ($r = 0.39$), and *strong* and *alarming* ($r = 0.38$). Directionally, subjects found rougher patterns more alarming and unpleasant. Stronger patterns were perceived as more alarming and rhythmic patterns were more pleasant ($r = 0.2$).

Stimulus composition influenced subjective ratings. On average, rhythm significantly impacted ratings for all scales, while frequency only impacted the *calm/alarming* ratings. To examine the effect of rhythm and frequency on ratings, we ran five separate within-subject ANOVA tests with each rating scale as the dependent factor and rhythm, and frequency as two independent factors. All reported effects were significant at $p < 0.01$. Rhythm had a main effect on all five scales (see Table 2.1). The long continuous vibration (pattern 1) was perceived as strongest, smoothest, and most non-rhythmic. The pattern with several very short vibrations (p6) was the roughest, most alarming and most unpleasant. The long vibration with one short silence (p4) was most pleasant and among the strongest. Patterns with few short vibrations (p3, p7) were the weakest and most calm. Frequency only had a main effect on the *calm/alarming* scale (Table 2.1). 175 Hz vibrations were more alarming than 75 Hz. There was an interaction effect of rhythm*frequency for *weak/strong* scale, i.e., 75 Hz was perceived stronger or weaker than 175 Hz depending on the pattern.

Individuals' affective and sensory ratings varied. The average ratio of mean to standard deviation for the five scales were: *weak/strong*: 0.71, *smooth/rough*: 0.27; *non-rhythmic/rhythmic*: 0.87; *calm/alarming*: 0.45; *unpleasant/pleasantness*: 0.22.

Table 2.1 Summarized results of the ANOVA tests on the five affective rating scales

Rating scale	Significant factors	F value, Effect size
Weak/Strong	Rhythm	$F(3.07, 107.44) = 49.46$, $\eta^2 = 0.58$
	Rhythm*Frequency	$F(6, 210) = 7.5$, $\eta^2 = 0.18$
Smooth/Rough	Rhythm	$F(2.8, 100.83) = 6.44$, $\eta^2 = 0.15$
Non-rhythmic/Rhythmic	Rhythm	$F(3.11, 112) = 25.94$, $\eta^2 = 0.42$
Calm/Alarming	Rhythm	$F(3, 109) = 10.64$, $\eta^2 = 0.23$
	Frequency	$F(1, 36) = 10.62$, $\eta^2 = 0.23$
Unpleasant/Pleasant	Rhythm	$F(2.75, 99) = 4.1$, $\eta^2 = 0.1$

Thus, reactions varied most for *unpleasant/pleasant*, *smooth/rough*, and *calm/alarming* respectively, two of which are affective dimensions.

Individuals deviated from overall affective/sensory scale correlations. Since examining the complex patterns of all correlations for each subject is a large task, as a first step we analyzed the correlations for one pair of scales (*pleasant* and *alarming*). Post-experiment comments had suggested differences in subjects' opinions for these two dimensions, making it a promising place to look for evidence that differences exist. Alarming and unpleasant ratings did not correlate for 11 subjects ($r < 0.35$ and non-significant), but were highly correlated for seven other subjects ($r > 0.7$ and significant). Such a large variation in affect justifies further examination. In future analysis, we will investigate the complex patterns of correlations among all dimensions; for example, MDS and factor analysis may better reveal the structures in individuals' ratings.

Variation in subjective ratings did not correspond to demographic or NFT. For each scale, we ran a between-subject ANOVA using the sum of ratings for that scale as the dependent variable. Gender (two levels), culture (two), music background (three), and NFT category (two) were the between-subject factors. We did not find a significant effect of these factors on the ratings. The effect size of NFT was very small (less than 0.1) which did not justify its practical significance even for a larger sample size.

Task performance varied, but variation did not coincide with affective ratings. Total score in each task, calculated as the sum of negative and positive scores for all items, varied from 50 to 85% for both tasks. However, all subjects performed above chance (>50% in the time task and >33% in the frequency task). Also, the distribution of our task scores did not show distinct groups of performance, in contrast to previous individual difference studies [7, 9, 10]. The distribution for the time task suggested three overlapping normal distributions which we used to divide subjects into three groups. The distribution for the frequency task was even more flat. For consistency, we divided subjects into three groups of low, medium and high scores (see Fig. 2.6); these groups held different members than for the time task. However, variations in subjective ratings did not correspond to time and frequency task performance in our study.

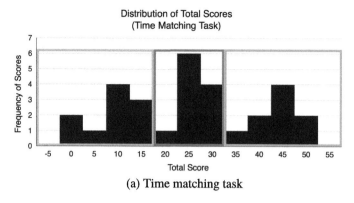

(a) Time matching task

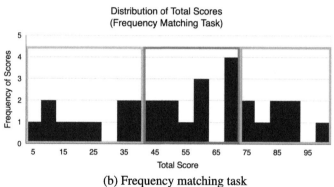

(b) Frequency matching task

Fig. 2.6 Distribution of total scores in time and frequency tasks; colored boxes show one possible grouping for the tasks

2.5 Discussion

We now relate our study results to our near-term research questions.

2.5.1 Dimensionality and Utility of Affective Response

2.5.1.1 What are the Relevant Dimensions for Measuring Affective Response, and is there Utility in Multiple Rating Dimensions?

We derived five affective and sensory dimensions for rating vibrations using literature and pilot studies (Sect. 2.3.4). Here we point to the findings that emerged from analyzing crosslinkages between affective and sensory dimensions.

Ratings showed a structure in affect and sensory ratings that might extend to other modalities. Based on the correlation among ratings, the vibrations were mostly perceived as *rough, alarming,* and *unpleasant*; or, *smooth, calm,* and *pleasant*. This organization can point to the inherent association of these attributes in subjects' mind. Future work can examine whether this structure holds for other vibrations and even other modalities.

Our stimulus set largely bypassed the positive valence/positive arousal region of the emotion response space. On average, few alarming vibrations received pleasant ratings. However, exciting rhythms (positive valence and arousal) are conceivable for vibrations and seem to be a relatively unexplored part in our vibrations. Thus, ratings on multiple dimensions can guide future stimuli design.

Affective and sensory ratings showed how individuals' patterns of preference deviated from average. Based on the correlation matrix for each subject, several subjects deviated from the overall correlation between unpleasant and alarming ratings. The integrated set of affective and sensory dimensions also enable investigation of more complex structures in future.

This initial set of scales needs further development and validation. As a first step, their utility in describing synthetic stimuli (e.g., various vibrations and tactile technologies) must be developed. Eventually, the proposed dimensions must evolve to support rating of natural stimuli, as a means to compare users' response to synthesized and natural stimuli. We also need to determine how accurately these dimensions can reflect human affective response in real-world contexts. One possibility is to test how well the rating instrument assists haptics designers in creating tactile stimuli that are indeed preferred by users in real-world scenarios. Another is to use neuroimaging studies to compare brain patterns for ratings to those for natural pleasant stimuli, e.g., fur.

2.5.2 Vibration Parameters

2.5.2.1 What Parameters From the Vibration Design Space Impact Affective Response, and How?

On average, rhythm pattern (duration of vibrations, number and timing of pauses) influenced subjective ratings for all five affective and sensory scales. Frequency only significantly impacted *calm/alarming*. Overall, rhythm pattern impacted the ratings the most. Drilling down: vibration duration directly influenced *weak/strong* ratings and the number of pauses determined *smooth/rough* and *calm/alarming* ratings. Overall, longer vibrations with fewer pauses were perceived as smooth and pleasant. Several short vibrations were considered rough, alarming and unpleasant.

The affective range in response to these vibrotactile stimuli is more limited than what we would expect to find for natural stimuli. However, even this small study found distinct preference for some vibrations over others. This suggests that having a scale can help designers now using this relatively inexpressive media in avoiding negative affect and designing more acceptable feedback. With improved rendering technology, we can expect to move towards more engaging touch sensations.

Some individuals' ratings diverged considerably from these overall trends, as indicated by the average ratio of mean ratings to standard deviation. Rating variations were especially high for *unpleasant/pleasant*, *smooth/rough*, and *calm/alarming* scales which were also highly correlated. In future, using a composite value based on ratings for the three dimensions might reveal different clusters of subjects and preferences.

2.5.3 Demographic, NFT Score and Tactile Performance

2.5.3.1 What is the Link Between Affective Response and Demographics, NFT Scores, and Tactile Task Performance?

Subjective ratings did not coincide with demographics, NFT scores, or tactile abilities. Our results are consistent with past studies which also did not find any considerable effect of demographics. Regarding NFT, we had determined *a priori* that 30 subjects would not have enough power to detect an effect (Sect. 2.2), but we included the NFT questionnaire to assess its sensitivity. Our results suggest a very small effect size for NFT (less than 0.1 on subjective ratings). Regardless of power of a later study, such a small effect on subjective ratings does not have practical significance. NFT might not be sensitive enough to account for the affective range of synthetic stimuli. We thus plan to exclude the NFT in future work with synthetic stimuli and focus on tactile performance. For natural stimuli with a larger range of affective response, NFT might prove a more useful instrument.

To assess our results for tactile performance, we need to answer two questions:

1. How well did the time and frequency tasks reflect tactile abilities? Our analysis suggested that the frequency task better reflected tactile abilities (reasonable validity and reliability) but the reliability of the time task needed improvement. First, both tasks had a reasonable difficulty level to generate a low to high performance range (50 to 85% of correctly matched items). Second, our analysis suggests that the tasks relied on tactile sensory memory (subjects' scores in the two tasks did not correlate with their report of using pitch or rhythm for matching the stimuli). As a future test of discriminant validity, we can compare subjects' performance in auditory versus tactile matching tasks. Finally, the correlation between the two rounds of the frequency task ($r = 0.67$) and the two rounds of the time task ($r = 0.37$) indicated a reasonable reliability for the frequency task, while the time task needed improvement. Convergent validity of the tasks must be established in future, e.g., by using time and frequency discrimination tasks.

2. Do individuals exhibit considerable differences in tactile processing ability? Although task score distributions showed some variations in performance, they did not suggest obvious groupings. In contrast, past studies reported distinct groups of performers. What was the reason for these different results? Are there real differences in people's tactile abilities? In retrospect, almost all studies reporting huge individual difference in task performance involve a spatial component [7, 9, 10]. So it could be that people are different in some aspects of tactile abilities and not in others. If so, a battery of tasks is needed to measure tactile abilities. Moreover, most of those past studies used a specific instrument (Optacon), and their tasks had a cognitive component involved: subjects needed to map a tactile pattern to its visual representation. Both the instrument characteristics and the cognitive element could cause the variations in performance. A next step would be to study the potential differences in spatial tactile tasks by eliminating those confounds.

Based on past work, we started with the hypothesis of considerable differences in tactile abilities; we did not see this in these particular conditions. Now, the question is: Do people vary substantially in their processing of tactile stimuli; if so, in what respect? Does learning account for those differences? Only after answering these questions we can examine links between tactile abilities and affective response.

2.6   Conclusion

In this chapter, we examined affective response to vibrations for a handheld device. We presented our progress towards an integrated set of rating scales for measuring various dimensions of affect and perception, specifically *weak/strong, smooth/rough, non-rhythmic/rhythmic, calm/alarming,* and *unpleasant/pleasant.* Using these scales, we measured subjective response to rhythm pattern and frequency of vibrations. The correlation of ratings indicated that subjects found smooth patterns and rhythmic patterns more pleasant. Rougher patterns as well as stronger vibrations were perceived more alarming. According to the overall ratings, pleasant and alarming vibrations were relatively underrepresented in our vibrations and can be explored further in future. Within-subject ANOVA on the subjective ratings showed a main effect of the rhythm on all five rating scales, a main effect of frequency on the *calm/alarming* ratings, and interaction of rhythm*frequency for the *weak/strong* scale. Ratings varied considerably among subjects for *unpleasant/pleasant, smooth/ rough,* and *calm/alarming* dimensions. However, demographics, NFT scores and task performance did not coincide with these variations.

This study highlights several directions for future research in this area: (1) *Measurement tools*: Do affective responses to naturalistic stimuli differ qualitatively from those to synthetic stimuli, like vibrations; and can the same assessment tools uncover both types of responses? (2) *Key Attributes*: To what extent the effects of rhythm and frequency generalize to other tactile technologies? What other signal parameters are affectively important? (3) *Individual Differences*: How can we quantify individuals' deviation from the overall patterns of ratings for affect and sensation? Can we cluster people based on these patterns? To what extent individuals vary in other tactile tasks, e.g., tactile spatial tasks? What is the role of learning?

Answering these questions not only provides a better picture of affect and perception of tactile sensations but can also guide development of a standardized set of affective rating scales for evaluating haptic stimuli.

References

1. Field, T.: Touch. MIT Press, Cambridge (2003)
2. Essick, G.K., McGlone, F., Dancer, C., Fabricant, D., Ragin, Y., Phillips, N., Jones, T., Guest, S.: Quantitative assessment of pleasant touch. Neurosci. Biobehav. Rev. **34**(2), 192–203 (2010). http://www.sciencedirect.com/science/article/pii/S0149763409000190
3. Swindells, C., MacLean, K.E., Booth, K.S., Meitner, M.: A case-study of affect measurement tools for physical user interface design. In: Proceedings of Graphics Interface (GI '06), pp. 243–250 (2006)
4. Takahashi, K., Mitsuhashi, H., Murata, K., Norieda, S., Watanabe, K.: Feelings of animacy and pleasantness from tactile stimulation: Effect of stimulus frequency and stimulated body part. In: 2011 IEEE International Conference on Systems, Man, and Cybernetics (SMC), pp. 3292–3297 (2011). http://ieeexplore.ieee.org/xpls/abs_all.jsp?arnumber=6084177
5. Nagano, H., Okamoto, S., Yamada, Y.: What appeals to human touch? effects of tactual factors and predictability of textures on affinity to textures. In: Proceedings of IEEE World Haptics Conference (WHC '11), pp. 203–208 (2011). http://ieeexplore.ieee.org/xpls/abs_all.jsp?arnumber=5945486
6. Zheng, Y., Morrell, J.B.: Haptic actuator design parameters that influence affect and attention. In: Proceedings of IEEE Haptics Symposium (HAPTICS '12), pp. 463–470. IEEE (2012)
7. Levesque, V., Oram, L., MacLean, K.E.: Exploring the design space of programmable friction for scrolling interactions. In: Proceedings of IEEE Haptic Symposium (HAPTICS '12), pp. 23–30 (2012)
8. Peck, J., Childers, T.L.: Individual differences in haptic information processing: the need for touch scale. J. Consum. Res. **30**(3), 430–442 (2003)
9. Cholewiak, R.W., Collins, A.A.: Individual differences in the vibrotactile perception of a "simple" pattern set. Atten., Percept., Psychophys. **59**(6), 850–866 (1997)
10. Epstein, W., Hughes, B., Schneider, S.L.: Perceptual learning of spatiotemporal events: evidence from an unfamiliar modality. J. Exp. Psychol.: Hum. Percept. Perform. **15**(1), 28 (1989). http://psycnet.apa.org/journals/xhp/15/1/28/
11. Artificial Muscles Inc.: ViviTouch - "Feel the game.". http://vivitouch.com/. http://vivitouch.com/. Accessed: 28 Sept 2012
12. Guest, S., Dessirier, J.M., Mehrabyan, A., McGlone, F., Essick, G., Gescheider, G., Fontana, A., Xiong, R., Ackerley, R., Blot, K.: The development and validation of sensory and emotional scales of touch perception. Atten., Percept., Psychophys. **73**(2), 531–550 (2011)
13. Ternes, D., Maclean, K.E.: Designing large sets of haptic icons with rhythm. In: Haptics: Perception, Devices and Scenarios, pp. 199–208. Springer (2008)
14. Hoggan, E., Raisamo, R., Brewster, S.A.: Mapping information to audio and tactile icons. In: Proceedings of the 2009 International Conference on Multimodal Interfaces (ICMI '09), pp. 327–334 (2009). http://dl.acm.org/citation.cfm?id=1647382
15. Lederman, S.J., Klatzky, R.L.: Haptic perception: a tutorial. Atten., Percept., Psychophys. **71**(7), 1439–1459 (2009)
16. Harris, J.A., Harris, I.M., Diamond, M.E.: The topography of tactile working memory. J. Neurosci. **21**(20), 8262–8269 (2001). http://www.jneurosci.org/content/21/20/8262.short
17. Kaas, A.L., Stoeckel, M.C., Goebel, R.: The neural bases of haptic working memory. In: Human Haptic Perception: Basics and Applications, pp. 113–129 (2008). http://www.springerlink.com/index/p373185430172257.pdf

18. Hollins, M., Bensmaïa, S., Karlof, K., Young, F.: Individual differences in perceptual space for tactile textures: evidence from multidimensional scaling. Percept. Psychophys. **62**(8), 1534–1544 (2000)
19. Biggs, S.J., Hitchcock, R.N.: Artificial muscle actuators for haptic displays: system design to match the dynamics and tactile sensitivity of the human fingerpad. In: Proceedings of SPIE, vol. 7642 (2010). http://spie.org/x648.html?product_id=847741
20. Orr, M.G., Ohlsson, S.: Relationship between complexity and liking as a function of expertise. Music. Percept. **22**(4), 583–611 (2005). http://www.jstor.org/stable/10.1525/mp.2005.22.4.583
21. Alter, A.L., Oppenheimer, D.M.: Uniting the tribes of fluency to form a metacognitive nation. Pers. Soc. Psychol. Rev. **13**(3), 219–235 (2009)
22. Ekstrom, R.B., French, J.W., Harman, H.H., Dermen, D.: Manual for kit of factor-referenced cognitive tests. Princeton, NJ: Educational Testing Service (1976). http://www.ets.org/Media/Research/pdf/Manual_for_Kit_of_Factor-Referenced_Cognitive_Tests.pdf
23. Busey, T.A., Tunnicliff, J., Loftus, G.R., Loftus, E.F.: Accounts of the confidence-accuracy relation in recognition memory. Psychon. Bull. Rev. **7**(1), 26–48 (2000). http://www.springerlink.com/index/V247323112X74X2G.pdf
24. Knowles, W.B., Sheridan, T.B.: The "Feel" of rotary controls: friction and inertia. Hum. Factors **8**(3), 209–215 (1966). http://hfs.sagepub.com/content/8/3/209.short

Chapter 3
Characterizing Personalization Mechanisms

Abstract In Chap. 2, we found that individual differences in affect cannot be simply modelled based on users' tactile performance or background. Here, we examine *personalization* as a way to leverage the affective qualities of vibrations and satisfy diverse tastes; specifically, the desirability and composition of vibrotactile personalization tools for end-users. A review of existing design and personalization tools (haptic and otherwise) yielded five parameters in which such tools can vary: (1) size of design space, (2) granularity of control, (3) provided design framework, (4) facilitated parameter(s), and (5) clarity of design alternatives. We varied these parameters within low-fidelity prototypes of three personalization tools, modeled in some respects on existing popular examples. Results of a Wizard-of-Oz study confirm users' general interest in customizing everyday vibrotactile signals. Although common in consumer devices, *choosing* from a list of presets was the least preferred, whereas an option allowing users to balance vibrotactile design control with convenience was favored. We report users' opinion of the three tools, and link our findings to the five characterizing parameters for personalization tools that we have proposed.

3.1 Introduction

Increasingly present in consumer electronics, vibrotactile stimuli generate mixed reactions. Genuine utility is possible, yet a given user may find the stimuli themselves unsuitable in their context, but cumbersome if not impossible to modify. A common example is call or message notifications in cellphones, generally provided with a limited set of basic vibrations (or perhaps just one) that cannot accommodate the broad range of user preferences.

This problem is not merely aesthetic: mappings between stimuli and their meanings can be hard to learn when mnemonic links are not apparent, and meanwhile users may wish to deploy salience (e.g., due to amplitude, duration and repetition) according to an intensely personal scheme. When mappings and salience do not work well for an individual, utility is overwhelmed by irritation; the signals are relegated to minimal roles or disabled altogether.

© Springer Nature Switzerland AG 2019   39
H. Seifi, *Personalizing Haptics*, Springer Series on Touch
and Haptic Systems, https://doi.org/10.1007/978-3-030-11379-7_3

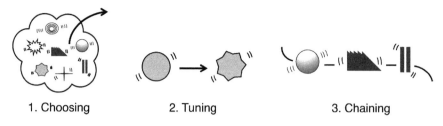

1. Choosing   **2. Tuning**   **3. Chaining**

Fig. 3.1 Conceptual sketch of three haptic personalization mechanisms

In this research, we are exploring the further premise that appropriately leveraging *affective* qualities of haptic stimuli in interface design could change this. Not only might "design for affect" add to the variety, pleasure and fun of using electronic devices, it could be exploited to enhance functional benefits by making individual signals more intelligible and memorable.

However, incorporating affect into haptic design is not easy. Affective responses to synthetic haptic stimuli are not yet well catalogued, precluding a heuristic approach at this time. Individual differences in both perception and affect further complicate the matter [1, 2]. While academic and industry experts are progressing towards a better understanding of affective response and design principles, we consider a different approach: *empower ordinary users, having no previous design knowledge, to design or personalize haptic feedback for their own preference and utilitarian needs.*

A first question is thus: *(Q1): What characteristics will make a vibrotactile personalization tool usable?* The design space for vibrotactile stimuli appears large if we consider all combinations of the controllable variables (e.g., frequency, amplitude, waveform and even rhythmic presentation). Yet, many are perceptually similar when rendered, and this further depends on device characteristics [3]. A typical user, with a limited conceptual model of this structure and its non-independence, would get little traction if given these comprehensive, low-level controls. Thus, we investigate the productivity and desirability of a diverse set of tools that might support typical end-users in personalizing haptic effects, with the dual hope of such utilities leading to better tools for haptics designers as well (Fig. 3.1).

The second question is whether given a manageable tool, this is desirable. Specifically, *(Q2) Do users* **want** *to personalize vibrations for their everyday devices?*

Finally, as a step towards understanding affective preferences themselves, we wonder *(Q3): What kind of vibrations do people design when given the opportunity?*

In this chapter, we focus on Q1, and establish insights and future directions for Q2 and Q3. We identified parameters that characterize existing personalization tools, then evaluated their manifestations in three haptic tool concepts via a Wizard-of-Oz (WoZ) study where we asked participants to design urgent and pleasant cellphone notifications (Fig. 3.2). Our contributions include:

- Five dimensions for vibrotactile design and personalization tools;
- Three tool concept prototypes that capture this variation;

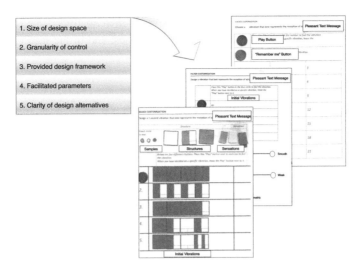

Fig. 3.2 Study paradigm: five proposed personalization tool parameters and three personalization tool concepts (low-fidelity prototypes) which capture variance in these parameters

- Quantitative and qualitative data on user opinions of the three concepts, viewed in context of the proposed tool parameters;
- Informal qualitative data on vibrations designed by users.

3.2 Related Work

3.2.1 Haptic Design

Haptic effects can take many forms, the most common of which is vibrotactile (also the focus of our work). By "haptic design", we refer to creating haptic effects to be rendered by a haptic display. Existing haptic devices vary considerably in their capabilities, leading to a tight coupling of effect design to device development. Haptic designers must intimately understand technical device parameters, and currently must usually design within that technical space. For example, vibrotactile designers can typically vary frequency, waveform, amplitude, duration and rhythm [3, 4]. Documentation of a mapping from technical space to users' perceptual space for tactile stimuli is underway [3, 5, 6]. Here, we have structured our proposed tools in an *intuitive and perceptual* rather than a *technical* control space, positing that this will lead to more satisfying results, particularly for inexperienced designers.

Vibrotactile effects have been designed both to communicate information (see [4] for a survey) and affect [7]. To ensure effective design, haptics designers typically use iterative design and user evaluation of haptic stimuli [4]. However, this approach has been less successful for haptic effects with affective qualities; convergence is difficult in the absence of adequate evaluation metrics, and in the face of notable individual preference differences (Chap. 2).

3.2.2   Haptic Design Tools

The haptic community has proposed a number of design tools in the past decade, each aiming to reduce technical knowledge required for design and thus opening the domain to a wider audience.

Categorization of Tools: Paneels et al. [8] categorizes haptic design tools based on their support for one versus multiple actuators; and type of representation: a direct signal (e.g., Haptic Icon Prototyper [9], and Immersion's Haptic Studio [10]) or an indirect, metaphor-based view (e.g., VibScoreEditor [11], TactiPed [8]). We find that this organization does not adequately differentiate tools for end-user personalization. For example, all of our prototypes use indirect representation and currently support one actuator, yet vary in other substantive ways.

Creation and Modification: All the tools we have seen are primarily concerned with creating haptic effects. For example, to create vibrations, Hong et al. [12] mapped user touch input (e.g., pressure, location) to amplitude and frequency, an approach found useful for prototyping and demonstration but not suitable for modification of effects. Other tools support both creation and modification of the effects. The Haptic Icon Prototyper provides more flexibility by allowing users to combine short haptic snippets in a sequential or parallel form along a timeline [9]; one of our three concepts (*chaining*) uses a similar approach. With a focus on creation and modification, all the above tools provide fine-grained control over stimuli. For a modification-only tool, the importance of various tool requirements can shift – for example, convenience might outweigh design control. Here, we are also primarily interested in modification or personalization of pre-existing templates, as it could be a more practical approach for users without design knowledge.

Audience: Existing tools differ in the design knowledge they require and thus usability for ordinary users. Some (e.g., VibScoreEditor, TactiPed) specifically target ordinary users; but despite their promising evaluations, they have remained in the academic domain. A notable exception is the iPhone tapping tool for creating customized vibrations for a user's contact list [13].

3.2.3   Challenges and Potentials of End-User Personalization

While these tools typically aim to be accessible to ordinary users, these users' ability to design has rarely been investigated. Oh and Findlater [14] studied custom gesture creation by this group, and found they were able to create a reasonable set of gestures but tended to focus on variations of familiar gestures. Personalization might suit at least some end-users better than creation, affording satisfaction instead of frustration.

We can gain insight from personalization literature in software engineering on factors involved in end-user personalization of software applications. Sense of control and identity, frequent usage, ease-of-use and ease-of-comprehension in tools allowing personalization engender takeup [15] while personalization is discouraged by lack of time or interest, and difficulty of personalization processes [16].

3.3 Conceptualization of Haptic Personalization Tools

As a first exploratory attempt to conceptualize haptic personalization tools, we examined, brainstormed and discussed characteristics of existing design tools in the haptic and other domains. As a result, we propose five parameters along which design and personalization tools can vary, including: (1) size of design space, (2) granularity of control, (3) provided design framework, (4) facilitated parameter(s), and (5) clarity of design alternatives (Table 3.1). We posit that these parameters can influence users' perception of flexibility and effort to design haptic effects and consequently, their preference and tool choice.

Although desirable, dependencies among the parameters make it infeasible to study the effect of each parameter in isolation or to examine users' opinions about all variations of the parameters in a meaningful study. Existing tools co-vary on many of these parameters and a realistic study would need to examine many together. Thus, we define three haptic personalization tool concepts that are considerably different, capture variations along all tool parameters, and are practically interesting. Our concept prototypes borrow from existing tools in haptic and photo editing domains.

3.3.1 Three Personalization Tools

We begin by describing our three proposed tool concepts, implemented as paper prototypes, then use these and existing tools to explain our proposed tool characterization parameters. We chose to evaluate manually operated low-fidelity prototypes because a tool concept can be implemented in various ways differing in interface elements or interaction style and we wanted to avoid reactions focused on those differences. In contrast, a paper prototype allows users to flexibly interact with the tool concept, thus we could obtain reactions focused on conceptual differences of the tools.

1. Choosing[1] *(baseline: minimal personalization, focuses on convenience)*: This tool models a conventional way of personalizing ringtones and other auditory alerts on consumer electronics, wherein users are provided with a list of vibrations to choose

---

[1]Called "choice" in the original conference publication.

 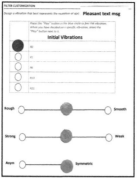 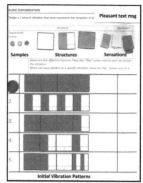

(a) *Choosing* concept: a 7x3 table of vibration pre-sets lies beneath blue *Play* and orange *Remember Me* buttons. In this paper prototype, moving the blue or orange sticker to one of the vibration cells represents (in a real device) cursor-selection of a vibration and then the execution of that function on it. In our WoZ study, the experimenter executed this response manually, s.t. the participant felt the selected vibration on the display device.

(b) *Tuning* concept: user can apply 3 filters (bottom) to 5 rhythm presets (top); the presets cannot otherwise change. The roughness and strength filters have three settings each, and the symmetry filter has two. The blue *Play* button again selects a preset. Here, the movable orange *Level* circles show the current filter settings for playback (shown: default setting).

(c) *Chaining* concept: lower area visualizes the time sequence for 5 initial rhythms (purple indicates vibration-on, and white is silence, over a 500ms period). Users can modify the rhythm itself by selecting and overlaying a different block structure (top middle) and an available block sensations (colored rectangles on top right). The 3 small colored circles (top left) allow users to try the 3 block sensations (45Hz, 75Hz, 175Hz) before using them.

Fig. 3.3   Three personalization tool concepts

from. Our prototype (Fig. 3.3a) lists the vibrations in a tabular structure where rhythm varies by row and vibrotactile frequency by column. The user places the *Play* button over each vibration number to signal to the experimenter (acting as a computer) to play the vibration. The *Remember Me* buttons are used to mark some vibrations and facilitate future comparison and choice.

2. Tuning[2] *(more power, still emphasizes convenience by allowing high level control)*: Inspired by color adjustment filters in photo editing tools like Adobe Photoshop, users have a small initial set of vibrations and three perceptual filters to vary roughness, strength, and symmetry. These dimensions have repeatedly emerged as the most salient and important [3]. *Tuning*'s paper prototype (Fig. 3.3b) includes five initial vibration patterns in the upper rows, and three sliders representing the filters at the bottom. To feel a vibration, users need to choose a rhythm at the top with a particular setting of the filters at the bottom.

---

[2]Called "filter" in the original conference publication.

3. Chaining[3] *(trades off convenience for greater control over the stimuli)*: Derived from the Haptic Icon Prototyper [9], a vibration is made of a sequence of vibration blocks and to modify a vibration, users change the individual blocks in the sequence using the available vibration blocks. With our prototype (Fig. 3.3c), users can start from one of the five vibrations at the bottom, then choose a block structure (silence, half vibration, and full vibration) and one of the three block sensations from the top and place it at the desired location along the chosen vibration sequence. They can test their design by putting the blue circle (*Play* button) beside the vibration.

3.3.2 Proposed Tool-Characterization Parameter Space

We were able to identify five parameters that described the variation we observed during our review of existing personalization tools. Table 3.1 relates these parameters to our three concept prototypes (*choosing*, *tuning* and *chaining* personalization). These parameters are not orthogonal or independent: for example, providing finer control over stimuli will increase the size of the design space.

(1) Size of Design Space Accessed by the Tool: The size of the design space refers to the number of distinct stimuli that a tool can create; it depends on the design tool and a rendering haptic display. The tool's "perceptual size", meaning the number of *perceptually distinct* stimuli that it can create, is also important but harder to quantify. For example, if people can only distinguish a subset of stimuli designed by a tool and rendered by an actuator, that subset is the perceptual space for that tool and actuator. The size of design space increases from *choosing* to *tuning* and to *chaining*.

(2) Granularity of Control: The smallest unit of a stimuli that a user can directly manipulate with a tool can vary from holistic (coarse) to local (fine) control. With *choosing* and *tuning*, users could control a whole 2 s vibration by selecting it, but

Table 3.1 Embodiment of proposed parameters: characterization of *choosing*, *tuning* and *chaining* concepts

Proposed parameters	Choosing	Tuning	Chaining
1. Size of design space (for C2 tactor [17])			
Technical:	21	90	2400
Perceptual:	21	$\sim 45 - 90$	< 2400
2. Granularity of control	Holistic (Coarse)	Holistic (Coarse)	Detailed (Fine)
3. Provided design framework	List	Perceptual	Building blocks, Outline
4. Facilitated parameter(s)	Feel, Rhythm	Feel	Rhythm
5. Visibility of alternatives	High	High	Low

---

[3]Called "block" in the original conference publication.

with *chaining* they had control over 125 ms sub-blocks (by modifying or replacing them).

(3) Provided Design Framework: Any design tool inevitably imposes an outline or framework on design. This structure will, to some degree, impose on the user some organization of the design space. Our *choosing* tool provides the tightest structure, by only allowing users to choose from a list of sorted vibrations. *Tuning* conveys a perceptual organization of the design space, via the three axes provided. *Chaining* provides a discrete, block-based outline for the design and organizes building blocks into 3 structures (rhythm management) and 3 sensations (frequencies). As another example, the iPhone tapping tool provides very little structure: vibrations are viewed as variable-length touches to the screen.

(4) Facilitated Parameters: The degree and ease of control that a given tool affords for each parameter may vary. Some are promoted by the tool for creation or manipulation of stimuli and take the least or little effort to manipulate. *Chaining* facilitates control over the rhythm or structure of vibration while *tuning* facilitates control of feel or sensation. Both of these tools to some extent allow control over structure and feel but one is more prominent than the other. *Choosing* allows limited control over both feel and rhythm.

(5) Visibility or Clarity of Design Alternatives: Tools vary on the extent that alternative designs are provided to users, versus discovered. Visibility of design alternatives decreases from *choosing* (all stimuli are listed) to *tuning* (all filter combinations are apparent) to *chaining* (outline and building blocks are apparent, many versions are possible. Traversal of the design space in a reasonable time must involve discovery).

3.4   Methods

We ran a WoZ study with paper prototypes to examine users' interest in personalization and their opinions of our tool concepts.

Setup: We delivered vibrotactile effects with a C2 tactor [17], controlled via a control computer's audio channel and audio-amplified; signal and amplification levels were held constant. To maximize dynamic range, participants held the actuator between the thumb and index finger of the dominant hand and worked with one prototype at a time (Fig. 3.4). They used movable paper pieces to specify vibrations; when they pressed the movable blue *Play* button, the experimenter played back those vibrations to them. Participants could not see the control laptop screen.

Stimuli: All vibrations in the study lasted 2 s. Vibration duration and other choices for the parameter values were determined based on pilot studies and prior work. We used 7 rhythm patterns (Fig. 3.5) from a larger rhythm set [3]. Initial vibrations and possible alternatives varied for each tool:

 1. *Choosing*: 7 rhythms (Fig. 3.5) were rendered in 3 frequencies (45 Hz, 75 Hz, 175 Hz), chosen based on pilot studies. Thus, participants could choose from a total of 21 vibrations arranged in a table: the vibrations with different rhythms in rows and those with different frequencies in columns (Fig. 3.3a).

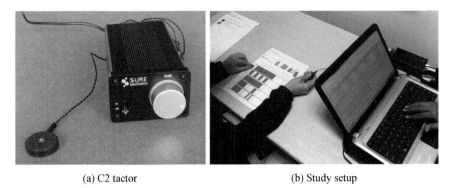

(a) C2 tactor   (b) Study setup

Fig. 3.4   Study setup showing a participant working with a prototype and the experimenter playing back the vibrations

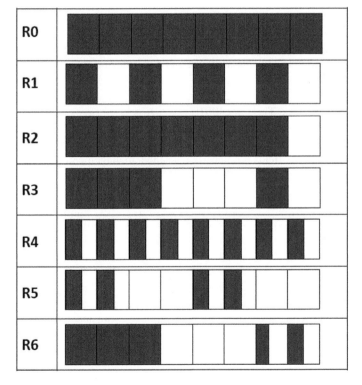

Fig. 3.5   Seven rhythm patterns: each row represents a vibration pattern which is repeated 4 times in a 2 s stimulus

Table 3.2 Configurations of each filter setting in the *tuning* tool

Setting	Change from default vibration
Default	No change (75 Hz, 5 first rhythms from Fig. 3.5)
Smooth	45 Hz, De-amplification of 3 dB
Rough	5 ms silence added to middle of each 50 ms vibration
Weak	De-amplification of 6 dB
Strong	Amplification of 6 dB
Asymmetric	Removal of 2/3rd of vibrations in the first second

2. *Tuning*: We rendered the first 5 rhythms in Fig. 3.5 in 75 Hz to represent the middle setting on the strength and roughness filters and the symmetric setting on the last filter. Participants could choose from 18 filter settings ($5 \times 18 = 90$). Entries of Table 3.2 show changes relative to the default settings, determined by pilot studies and prior work in our group to match the perceptual filter labels.

3. *Chaining*: The first 5 rhythms in Fig. 3.5 were initial templates for *chaining* personalization. To make a new vibration, one could choose one of the 3 block *structures* (silence, half vibration, and full vibration) with one of the 3 block *sensations* (45 Hz, 75 Hz, 175 Hz). Each block had 125 ms duration; the full pattern was 500 ms, to be repeated 4x in playback. This left 2400 ([2 *vibration structures* $\times$ 3 *sensations* + 1 *silence structure*]^4 − 1) design alternatives.

Participants: 24 university students (9 male) participated in a 1 hour study for $10. They came from many fields (engineering, science, management, arts, etc.) and age range (16 [19–29 years], 4 [30–39], 3 [40–49], 1 [>50]). 20 used cellphones or game controllers with haptic feedback on a daily basis. 7 had basic design experience with Photoshop and other video editing software.

Design: We used one independent within-subject factor (prototype, three levels) and counterbalanced order of interface with a Latin square. We also counterbalanced order of designing urgent versus pleasant notifications, though for each participant, kept the order the same across the three prototypes. We collected: (1) ratings on personalization interest (1–5 Likert scale), (2) rankings of the tools on ease-of-use, design control, and preference, (3) comments from participants, (4) time spent on each tool, (5) vibrations designed with each tool for pleasant and urgent notifications.

Procedures: Study sessions took place in a quiet room. Participants completed a questionnaire on demographics, experience with haptic feedback, and previous haptic, auditory or visual design experience. The experimenter then briefly explained the first prototype and asked the participant to use it to design an urgent and a pleasant notification; repeated this for each tool (about 15 min each); and administered the post-questionnaire above. We also asked which tools they would use if they had all three tools on their cellphone and for what purpose; if they had enough time to design vibrations, and if the labels in the *tuning* tool matched the vibrations.

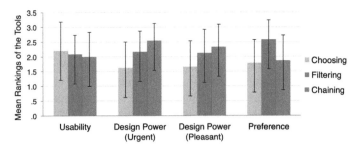

Fig. 3.6 Participants' rankings of the three tools. *Chaining* was the most powerful while *tuning* was the most preferred

3.5 Results

3.5.1 Comparison of the Tools

We use separate Friedman tests to compare the rankings of the tools on ease-of-use, design control and preference (Fig. 3.6). In the cases of statistical significance, we report follow-up pairwise comparisons using a Wilcoxon test and controlling for the Type I errors across these comparisons at the 0.017 level, using the Bonferroni correction.

Ease-of-Use or Usability: Ranking of ease-of-use did not differ significantly across the three interfaces ($\chi^2(2) = 0.8$, $p = 0.67$), suggesting that the usability of the tools were reasonably similar.

Design Control: Participants ranked how well each tool allowed design of an urgent and of a pleasant notification. There was a significant difference of interface for both types of messages (urgent: $\chi^2(2) = 10.94$, $p = 0.004$, pleasant: $\chi^2(2) = 6.02$, $p = 0.049$). For both types of messages, post-hoc tests indicated that *chaining* was significantly ranked more powerful than *choosing*, (urgent: $p = 0.003$, pleasant: $p = 0.041$). Rankings for *tuning* did not significantly differ from *chaining* and *choosing* (urgent and pleasant $p > 0.5$).

Preference: Rankings for preference were significantly different for the tools ($\chi^2(2) = 9.69$, $p = 0.008$). Post-hoc comparisons showed *tuning* was significantly preferred over *choosing* ($p = 0.006$) and *chaining* ($p = 0.012$).

Design Time: According to the post-questionnaire, participants generally had enough time; three participants wanted more time for *chaining*, the most complex. The average time spent on *chaining* ($M{\sim}12.5$ m, $SD{\sim}5$ m) was higher than for *tuning* ($M{\sim}7$, $SD{\sim}2.5$) and *choosing* ($M{\sim}6$, $SD{\sim}2.5$). This time included creation and playback of the vibrations by the experimenter. As we knew that vibration creation was more time-consuming for *chaining*, we did not analyze the timing data statistically. Our observations during the study sessions support the timing data i.e., participants needed more time to think, change, and compare the generated vibrations with *chaining*.

Choice of Tools: In response to our question "Which tools would you use if you had all three tools on your cellphone?", 20 participants (83%) chose *tuning*, 10 chose *chaining* (42%), and 8 chose *choosing* (33%). Unsurprisingly, many participants mentioned design flexibility and required time as two factors in their decision. According to their comments, *tuning* is *"simple and fast...yet gives flexibility to choose and customize" (P16)*. Interestingly, some participants described *chaining* as being *"fun" (P11)*, or for when they are in a *"good mood" (P15)*: *"When I feel that I have too much time and have a good mood, I may like to design a special pattern using the chaining personalization. If I don't have any mood or feel lazy, I may use the choosing or the tuning one."(P15)*

A majority (20/24) felt that the filter labels in *tuning* personalization matched the sensations. Three said that asymmetric and symmetric vibrations were not very different and one had a similar comment for the strength and roughness filters.

When we asked about the iPhone tapping tool, only three participants had tried it for making custom vibrations, none of whom found it useful. P24 doubted his/her ability to make nice vibrations: *"At first, I thought it would be fun making your own custom vibration, but once I tried the interface, I was not really into it since the vibrations I created were not as nice as the already customized vibrations on my phone."*

P9 wanted some vibration or structure to start from: *"It's simple and not so much patterns to choose from."*

P5 did not find the input mechanism adequate for his/her needs: *"It was really easy to use, but my fingers don't move fast enough to create the rapid vibration I would want to use for urgent messages. And it was hard to make the vibration symmetrical."*

3.5.2 Interest in Personalization

On average, participants stated mild interest in personalizing their vibration notifications ($M = 3.42$, $SD = 1.14$ on a 1–5 Likert scale). Lack or minimal use of vibrations was the main reason for not being interested in personalization while recognizing different types of alerts, being unique, adjusting the sensation levels, and concerns about repetitive exposure to unpleasant vibrations were the main reasons for personalizing their cellphone notifications.

3.5.3 Vibrations Designed by Participants

24 participant each designed 6 vibrations (one pleasant and one urgent with each tool) resulting in 144 in total. We provide an informal summary of the vibrations. We imagine that participants might have made different choices if designing for real use, and the WoZ study approach could also have impacted the extent that they explored

alternative designs. This might also be the reason for some inconsistencies in the vibrations designed with the three tools.

Overall, participants chose and modified the first three rhythms (R0, R1, and R2 in Fig. 3.5) the most. The order of rhythms on the paper prototypes was the same for all participants and all interfaces. Although this result can be partially due to the presentation order, the same rhythm preferences stood out in another experiment (Chap. 2). Unexpectedly, in many cases participants did not choose markedly different rhythms for pleasant and urgent messages. We are interested in knowing if a similar pattern of choices would hold in real life.

With *choosing* and *chaining*, over 20 participants (83%) used higher or the same frequency for urgent notification than for pleasant notifications. With *tuning*, over 17 participants (70%) used the strong and symmetric settings for both pleasant and urgent messages. The participants varied the rough/smooth and rhythm settings the most to differentiate pleasant and urgent messages. Only 8 participants (33%) used the asymmetric setting, and 5 of them used it only for urgent notifications.

3.6 Discussion

3.6.1 Desirable Characteristics (Q1)

Not surprisingly, perception of design flexibility and low effort are the main factors in participants' choices.

Design space accessed and flexibility afforded by tool framework impacts users' perception of *Design Control*. The perceived size of the design space is larger for *chaining*. Also, *chaining* only provides building blocks for designing vibrations, and thus affords a more flexible structure compared to *tuning* and *choosing*. According to the rankings, *tuning* provides reasonable design control (not significantly lower than *chaining*) and *choosing* has the least design control.

Holistic control over stimuli and visibility of design alternatives can reduce the perception of *Effort*. On average, participants took much less time with *choosing* and *tuning* compared to *chaining*. Also, post-questionnaire comments from participants indicate that they perceived *tuning* and *choosing* faster and easier than *chaining*. Control granularity and visibility of design alternatives appear to contribute to perceived effort; these parameters were similar for *choosing* and *tuning* but different for *chaining*.

***Preference* is a function of the perceived *Design Control*, *Effort*, and *Fun*.** The *choosing* personalization, which is the most common tool for customizing sound and visual effects in consumer devices, was the least preferred option in our study as it provides minimal sense of control and flexibility. The participants found *chaining* time-consuming but *tuning* provided enough design control (not significantly different from *chaining*) and required little effort. Thus, it was preferred the most. Also, many found its perceptual structure of the design space intuitive and convenient. Also, we hypothesize that a low ratio of perceptual to actual size of the design space could cause disappointment, since many efforts could eventually feel similar. In *tuning*, these two sizes were very close (ratio~1) compared to *chaining*.

Some participants described *chaining* as fun, suitable for when they are in a good mood; i.e., gamelike. *Chaining*'s "Fun" may arise from a sense of discovery due to its less structured design alternatives.

Finally, we note that tools such as the iPhone tapping tool provide very little structure for users. Comments suggest that ordinary users (in contrast to designers) prefer some degree of structure and outline to restrict the design space and guide their design. P9 specifically stated that *"It (iPhone tapping tool) is simple, and not so much patterns to choose from"*.

3.6.2   Value and Outcomes (Q2, Q3)

Do users *want* to personalize vibrations? Overall, users registered interest in personalizing their notifications and playing with personalization tools on their mobile devices (Q2). The majority did not require detailed, fine control and preferred quicker holistic changes with more perceptual impact. Factors that typically impact software personalization behavior also appear to hold for haptics, including extent of usage, sense of control and identity, required time, and ease-of-use and comprehension of personalization tools. Other factors such as creativity, fun and available sensations could be more specific to personalizing stimuli. To further address this question, we need to investigate various everyday scenarios for using vibrations and survey users' interest in personalizing vibrations in each case.

What do users create or choose? Fully categorizing what people choose when given the opportunity (Q3) will be a major, and context-dependent endeavor. As a start, we found some general trends, such as associating urgency to signal energy and preference for some rhythms which are consistent with prior work (Chap. 2). However, the designed vibrations vary not only across individuals but also in some cases across the tools which is very likely due, at least partially, to our lab-based WoZ approach. A longitudinal study with the developed tools can provide a more comprehensive answer to this question.

3.6.3   Wizard-of-Oz Approach

Following our goal of focusing on personalization concepts with the low-fidelity prototypes, our WoZ prototypes and evaluation appeared to elicit natural feedback in most cases. Nonetheless, it is possible that the unrealistic delay between indicating a command and feeling the sensations skewed certain data; specifically, making it difficult for the participants to compare urgency and pleasantness. However, the impact of this on *tool* preference should be minimal. Participant questionnaire responses suggest that they understood and responded to the paradigm for each tool. *"[I prefer] tuning for first time exploring [the] available or default choices...[and] chaining for advanced personalization"(P20).* Further, this delay should negatively impact the preference for *chaining* as it had the greatest delay; but despite this, many rated *chaining* as their first or second choices.

3.7 Conclusion

In this chapter, we examined the desirability and practicality of personalizing everyday vibrations by ordinary users. We proposed five parameters that can impact users' perception of personalization tools including: (1) size of design space, (2) granularity of control, (3) provided design framework, (4) facilitated parameter(s), and (5) clarity of design alternatives. We used cellphone message notification as an example application and prototyped three concepts varying in these parameters, namely, *choosing*, *tuning* and *chaining* personalization.

Overall, our participants showed interest in personalizing vibrotactile effects. According to the results of a WoZ study, all three tools were reasonably usable. The participants preferred *tuning* over both *choosing* (current practice) and *chaining* because it provides some degree of design control but requires little design effort. *Chaining* personalization was the most demanding of time and effort but also the most powerful. Despite almost unanimous preference for the *tuning* interface, our results indicate that individuals' weights for design control, effort, and fun of a tool is different. Thus, an effective personalization tool needs to incorporate a suite of easy-to-use tools with different design controls and affordances to accommodate diverse personalization needs.

We did not conduct controlled studies to examine the effect of each parameter in isolation, since the parameters are not orthogonal and all combinations of them are not practically interesting. Instead, we defined three practical personalization tool concepts to capture the variability along those parameters. The proposed parameters were useful in understanding users' opinions of our tools and the iPhone tool. We think the actual size of the design space and flexibility of the design framework impacts perception of design control. Holistic control over stimuli and visibility of design alternative can reduce the perception of effort. Preference is a function of the perceived design control, effort, and fun of the interface.

Ongoing questions are whether our proposed parameters can adequately characterize new personalization approaches and their use for other scenarios as well as users' reactions to them; if there is an optimal subset of the parameters for characterizing the tools, and even a single optimal set of parameter values. These merit further study; however, we predict the last will be unproductive. Instead, we encourage tool designers to consider variations of their tools along these parameters to find the best parameter combination for their case, and to consider diversity in user preferences.

In terms of easing the personalization task, we see two immediate opportunities. The first is to use filters for stylizing or branding haptic effects, an approach used extensively in photo editing software and preferred by our participants. What properties do users want to change (e.g., emotion, sensation, or physical properties)? How much does it depend on the design case? How can one design an emotion or sensation filter? The second is to gamify design. Some participants thought using *chaining* was fun. We do not know of any haptic design games; these could increase interest in haptics and lead to crowd-sourced designs.

At minimum, intuitive end-user tools will allow professional designers to employ participatory practices. More inclusive tools and processes will expose users' criteria and desires for haptic effects, which is a significant current challenge in professional haptic design.

References

1. Peck, J., Childers, T.L.: Individual differences in haptic information processing: the need for touch scale. J. Consum. Res. **30**(3), 430–442 (2003)
2. Levesque, V., Oram, L., MacLean, K.E.: Exploring the design space of programmable friction for scrolling interactions. In: Proceedings of IEEE Haptic Symposium (HAPTICS '12), pp. 23–30 (2012)
3. Ternes, D., Maclean, K.E.: Designing large sets of haptic icons with rhythm. In: Haptics: Perception, Devices and Scenarios, pp. 199–208. Springer (2008)
4. MacLean, K.E.: Foundations of transparency in tactile information design. IEEE Trans. Haptics (ToH) **1**(2), 84–95 (2008)
5. Brown, L.M., Brewster, S.A., Purchase, H.C.: A first investigation into the effectiveness of tactons. In: Proceedings of World Haptics Conference (WHC'05), pp. 167–176. http://ieeexplore.ieee.org/xpls/abs_all.jsp?arnumber=1406930 (2005)
6. van Erp, J.B., Spapé, M.M.: Distilling the underlying dimensions of tactile melodies. In: Proceedings of Eurohaptics Conference, vol. 2003, pp.111–120 (2003)
7. Changeon, G., Graeff, D., Anastassova, M., Lozada, J.: Tactile emotions: A vibrotactile tactile gamepad for transmitting emotional messages to children with autism. In: Haptics: Perception, Devices, Mobility, and Communication, pp. 79–90. Springer. http://link.springer.com/chapter/10.1007/978-3-642-31401-8_8 (2012)
8. PanÃñels, S., Anastassova, M., Brunet, L.: TactiPEd: easy prototyping of tactile patterns. In: Proceedings of Human-Computer Interaction (INTERACT '13), pp. 228–245. Springer. http://link.springer.com/chapter/10.1007/978-3-642-40480-1_15 (2013)
9. Swindells, C., Maksakov, E., MacLean, K.E., Chung, V.: The role of prototyping tools for haptic behavior design. In: Proceedings of 14th Symposium on Haptic Interfaces for Virtual Environment and Teleoperator Systems, pp. 161–168. http://ieeexplore.ieee.org/xpls/abs_all.jsp?arnumber=1627084 (2006)
10. Immersion Corporation: Haptic Studio, http://www2.immersion.com/developers/. http://www2.immersion.com/developers/index.php?option=com_content&view=article&id=507&Itemid=835. Accessed 19 Sept 2013
11. Lee, J., Choi, S.: Evaluation of vibrotactile pattern design using vibrotactile score. In: Proceedings of IEEE Haptics Symposium (HAPTICS '12), pp. 231–238. http://ieeexplore.ieee.org/xpls/abs_all.jsp?arnumber=6183796 (2012)
12. Hong, K., Lee, J., Choi, S.: Demonstration-based vibrotactile pattern authoring. In: Proceedings of the Seventh International Conference on Tangible, Embedded and Embodied Interaction (TEI '13), pp. 219–222 (2013)
13. Wawro, A.: How to use custom vibrations in iOS 5 | PCWorld. http://www.pcworld.com/article/242238/how_to_use_custom_vibrations_in_ios_5.html. Accessed 24 Sept 2013

14. Oh, U., Findlater, L.: The challenges and potential of end-user gesture customization. In: Proceedings of ACM SIGCHI Conference on Human Factors in Computing Systems (CHI '13), pp. 1129–1138. http://dl.acm.org/citation.cfm?id=2466145 (2013)
15. Marathe, S., Sundar, S.S.: What drives customization?: control or identity? In: Proceedings of ACM SIGCHI Conference on Human Factors in Computing Systems (CHI '11), pp. 781–790. http://dl.acm.org/citation.cfm?id=1979056 (2011)
16. Mackay, W.E.: Triggers and barriers to customizing software. In: Proceedings of ACM SIGCHI conference on Human Factors in Computing Systems (CHI '91), pp. 153–160. http://dl.acm.org/citation.cfm?id=108867 (1991)
17. Engineering Acoustics, Inc.: C2 tactor https://www.eaiinfo.com/. https://www.eaiinfo.com/tactor-info/. Accessed 21 Mar 2017

Chapter 4
Choosing From a Large Library Using Facets

Abstract Choosing from example sets is a common mechanism for end-user personalization in many domains (e.g., colors, ringtones), but haptic collections are notoriously difficult to explore. This chapter addresses the provision of easy and efficient access to large, diverse sets of vibrotactile stimuli, on the premise that multiple access pathways facilitate discovery and engagement. We propose and examine five disparate organization schemes (facets), describe how we created a 120-item library with diverse functional and affective characteristics, and present *VibViz*, an interactive tool for end-user library navigation and our own investigation of how different facets can assist navigation. An exploratory user study with and of *VibViz* suggests that most users gravitate towards an organization based on sensory and emotional terms, but also exposes rich variations in their navigation patterns and insights into the basis of effective haptic library navigation.

4.1 Introduction

Vibrotactile technology appeared in mainstream consumer culture over a decade ago, first in buzzing pagers, cell phones, and game controllers. However, despite improvement in quality and expressiveness of consumer-grade tactile display, user appreciation and adoption has remained low.

One culprit is slow growth in the value added by haptics, e.g., "informative" uses wherein different stimuli have different assigned meanings [1, 2]. Low utility interacts closely with low liking: whether a user finds a tool hard to use or just dislikes it, he/she often responds to the consequent irritations, learning difficulty and incomprehensibility by minimizing or disabling it. The high incidence of online user posts for haptic features asking how to 'turn it off' suggests one or both of these are in fact happening with haptics.

Individual differences in haptic perception and preferences may be at the root of this problem. Underscoring this premise is the emerging theme of a need to recognize user diversity in end-user haptics research [3–7]. Would "turn-it-off" individuals see more value in tactile feedback if it met their *own* specifications?

© Springer Nature Switzerland AG 2019

H. Seifi, *Personalizing Haptics*, Springer Series on Touch
and Haptic Systems, https://doi.org/10.1007/978-3-030-11379-7_4

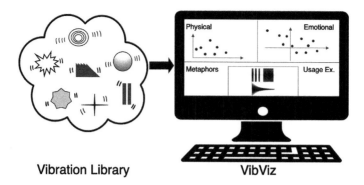

Fig. 4.1 Conceptual sketch of the *choosing* mechanism with *VibViz*

Diverse example sets, or libraries, are an obvious way to assist a user with personalization [4, 5]; but now we face the *navigation* challenge. Unlike visual images, vibrations must be scanned serially with most displays. Feeling and finding the entire contents of a sizable library is tedious and physiologically infeasible, as the first few vibrations quickly numb tactile receptors. Users may want to compare or choose multiple stimuli for their applications, but comparing and selecting from a rich multidimensional set is daunting. Confused and exhausted, users soon give up.

We are inspired by approaches taken in other domains to achieve highly navigable access to large, diverse collections. This includes principles such as offering multiple organizational schemes, informative and distinct visual representations, highlighting adjacencies between items and engaging users. While some publicly available vibrotactile libraries exist, the accessibility of this valuable resource is obstructed by the general absence of these elements.

Approach: The present research explores *how organization and representation of a vibrotactile collection can best support users in finding their desired vibrations*. Specifically, we identified five potential ways ("facets") for organizing effects. We created a library of 120 vibrations (for a single actuator), large enough to pose significant navigational needs, annotated it by the facets, and created *VibViz*, an interactive visualization interface with the goals of supporting both end-user navigation and our investigation of our five facets' utility and engaging qualities (Fig. 4.1). Finally, we conducted a preliminary evaluation of *VibViz* and the five facets using our vibrotactile library, in a user study with 12 participants where we triangulated questionnaire and observation data. Our contributions include:

- a process for creating a large (120 items) vibrotactile library
- identified challenges for large tactile library design
- five potential organization schemes (facets) for vibrotactile effects, drawn from literature
- an interactive library navigation interface (*VibViz*)
- a first evaluation of *VibViz* and the five facets

4.2 Related Work

4.2.1 Vibrotactile Libraries

Some large collections of vibrotactile effects exist, including Haptic Effects preview and Haptic Muse by Immersion (124 vibrations) [6, 7], and FeelEffects by Disney Research (>50 vibrations for a haptic seat pad) [5]. Each uses a single organizing principle: FeelEffects are grouped into 6 types of sensations or metaphors (e.g., rain, travel, motor sounds) and Haptic Muse by gaming use cases (sports, casino).

Other examples organize items on multiple dimensions simultaneously, but these axes occupy the same domain; e.g., van Erp (59 vibrotactile melodies) [8] and Ternes & MacLean (84 items varying on note length, rhythm, frequency, and amplitude) [9]. Relevantly, Ternes used MDS to translate a purely physical design space into perceptual dimensions [9], to facilitate "spacing out" its elements for maximum perceptual diversity given a device's capabilities.

Here we further hypothesize that restructuring a library over different *domains* will not only help optimize perceptual item packing for a given hardware's expressive capability, but also make it more accessible via multiple, qualitatively different means of exploring and understanding it.

4.2.2 Vibrotactile Facets

Vibrotactile effects can vary in many ways. Most examined are physical characteristics, including intensity, duration, temporal onset, rhythm structure, rhythm evenness, note length, and location [10], all measurable from the vibration signal. Research on tactile language suggests that users often describe vibrations with sensory and emotional words [8, 11, 12], motivating Guest et al.'s sensory and emotional dictionary for tactile sensations [13]. Schneider and MacLean found that people use familiar examples or metaphors (e.g., whistle, cat pawing) for describing vibrations [14]. Vibrations may also be characterized by their usage context (e.g., double click vibrations [6]) and example (cellphone vibrations).

We synthesized the above literature into five initial facets for vibrotactile effects, intended for structuring and accessing a large vibrotactile collection: (1) *Physical* characteristics—e.g., duration, energy ("1 second long"), (2) *Sensory* characteristics —e.g., roughness ("feels rough or changing"), (3) *Emotional* characteristics—e.g., pleasantness, arousal, and other emotion words ("feels urgent"), (4) *Usage Examples*—types of events for which a stimulus could be used ("good for a reminder"), and (5) *Metaphors*—familiar examples that resemble the effect in some way ("feels like snoring").

4.2.3   Inspiration From Visualization and Media Collections

Research on books and other media suggests that multiple visual pathways to a library can promote exploration and engagement, and increase serendipitous discovery [15]. Musicovery, an online music streaming service, visualizes its collection based on music mood and emotional content and allows filtering by genre, date, artist and activity [16]. However, unlike books and music, the most relevant alternative facets for vibrotactile stimuli have not been clearly identified.

Our library interface borrows many guidelines from the information visualization (InfoVis) domain, including using multiple views and linking their content. In InfoVis terminology, "filtering" refers to reducing the number of elements shown on the screen to a smaller subset of interest and a "glyph" can refer to any complex visual item, in contrast to single geometric primitives such as dots and squares [17].

4.3   Library and Facet Construction

Our library includes 120 vibrations, a size chosen to require an effective organization scheme. Elements range from 0.1 to 14.6 s in duration and 0.05 to 0.734 in energy (vibration signal Root Mean Square or RMS). In the present study, stimuli are rendered by a C2 actuator [18]. In the following, we describe how we designed the library and specified our five facets, and discuss obstacles we encountered.

4.3.1   Library Population

Our library required significant and diverse representation across all of our eventual facets to the extent possible given available physical parameters. We "sourced" effects through a variety of methods, including:

- collected a repository of effects from our past studies and collaborations with industry,
- systematically generated a large set of vibrations by varying the rhythm, frequency, and envelope structure,
- asked our haptics colleagues to design vibrations for a given list of metaphors (e.g., a dog, a spring, panting) with a rapid prototyping tool called mHive [12],
- constructed vibrations based on the Apple iPhone's sound icons, either mimicking timing and frequency changes, or directly applying low-pass filtering to them.
- for all of above, iteratively generated variants on existing vibrations and pruned overly-similar instances.

To balance facet representation, at several points we annotated the library's contents according to the current description of our facets. This in turn led us to refine our facet descriptions, with the final result in Table 4.1.

Table 4.1 Final vibrotactile facets used in study

1. Physical: Properties of a vibration that can be measured
(1) *duration* (msec), (2) *energy* (RMS), (3) *tempo* or speed (annotator-rated), (4) *rhythm structure*
For (4), we categorized stimuli by rhythm following [9]:
(a) *short note:* all pulses <0.25 s
(b) *medium note:* all pulses 0.25 s < 0.75 s
(c) *long note:* all pulses >0.75 s
(d) *varied note:* combination of short, medium, and long pulses
(e) *constant:* single pulse
2. Sensory: Vibration perceptual properties
(1) *roughness,* (2) *sensory words* from touch dictionary [13]
3. Emotional: Emotional interpretations of vibration
(1) *pleasantness,* (2) *arousal,* (3) dictionary *emotion words* [13]
4. Usage Examples: Types of events which a vibration fits
We collected and consolidated a set of usage examples for presentation timing and exercise tracking (Tam et al. [3])
5. Metaphor: Familiar examples resembling the vibration's feel
With a questionnaire, we collected a set of metaphors for our list of usage examples, asked colleagues and friends to provide metaphors for our vibrotactile effects, and used the NounProject website [19] for brainstorming on metaphors

4.3.2 Visualizing and Managing Diversity During Growth

As the library grew, it became harder to assess progress towards a goal of evenly distributed diversity; to compare existing effects, prune similar ones, and find gaps. We responded with several organization and visualization mechanisms.

(1) We built a database of existing vibrations in a spreadsheet; each row represented one vibration. Columns indicated vibration properties for each facet, and could be filtered. Despite addressing our most immediate needs, this approach had several drawbacks including limited filtering functionality, slow vibration playback, lack of a visual representation for the vibration patterns to support quick visual scanning.

(2) To improve visual inspection, we stacked subsets (about 30) of vibration waveforms in Audacity, an audio authoring tool, for quick vibrotactile modification and playback [20] (Fig. 4.2). The improved visualization qualities eased identification of near-duplicates and omitted vibration structures.

(3) Finally, we plotted vibrations according to their emotional (pleasantness and arousal) and physical characteristics (energy, duration, tempo, etc.) to enable successive pruning and filling along each dimension.

These mechanisms eventually conveyed us to an adequate result, but were cumbersome; worse, their fragmented nature hindered iteration, sometimes guiding modifications in conflicting directions. However, the experience of building this library

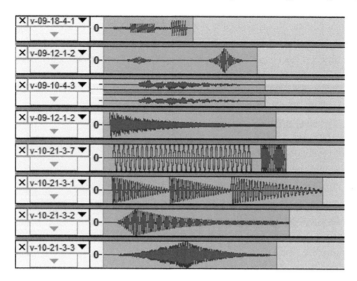

Fig. 4.2 Using Audacity for visual comparison of vibrations

gave us direct insight into the situation faced by any user in navigating a large, unstructured and poorly visualized set of items. The specific problem of navigation emerged as a primary obstacle to its use, whether for personalization or any other kind of design, and inspired us to turn to other interactive visualization mediums to craft a better solution.

4.4 VibViz: An Interactive Library Navigation Tool

4.4.1 Requirements

We needed our library interface to do two jobs, in the context of personalization tasks: (1) support novice end-users in vibration discovery (for example, in an online or local vibration library); and (2) allow us to study the utility and appeal of our five vibrotactile facets.

To support end-users, the interface must be easy to use without training. It needs to support both search and exploration; we anticipate that sometimes users will want to search with a set of characteristics in mind, and other times explore with minimal direction. It must support discovery of vibrations that resemble or contrast to a reference. It should provide multiple pathways, a key to serendipitous discoveries; and its use should be engaging enough to invite curiosity-driven exploration [15]. As a research tool, the interface needed to provide clear separation of the facets, allowing us to study user interactions by facet and users to articulate their opinions.

Table 4.2 *VibViz* user interface view descriptions

General Characteristics:

– Views A, B and C occupy the upper left, upper right and lower regions of the interface screen, respectively (Fig. 6.2a)

– We combined *Sensory* and *Emotional* facets due to tag overlap (View B). *Metaphor* and *Usage Example* facets share the vibration glyph on View C to save screen space

– Hovering over a dot (Views A-B) or row (View C) shows a visual thumbnail of the vibration pattern (glyph) and plays the vibration on the tactile display

A. Physical View: Provides an overview of all the vibrations, each represented by a coloured dot, according to axes of *energy* (vertical) and *duration* (horizontal)

Filters: (1) *Tempo*—slider for speed

(2) *Pulse structure*—checkboxes, with colours matching associated dots, for *short note, medium note*, etc.

(3) *Horizontal zooming*—click & drag on the Physical space zooms on the horizontal *duration* axis

B. Sensory and Emotional View: Each vibration appears as a dot in a 2D arousal—pleasantness space

Filters: (1) *Roughness* slider and (2) *Sensory and Emotion words* tagcloud. Changing the roughness range or clicking on the tagcloud selects vibrations having a roughness level in the specified range, and all of the currently selected tags

C. Metaphor and Usage Example View: A central, scrollable list of vibrations is flanked by *Metaphor* and *Usage Example* tagclouds. Each row has three columns: the vibration's *Metaphor* tags, its glyph, and its *Usage Examples*

Filters: Clicking on tags in either tagcloud reduces the displayed list to vibrations that have the specified tag(s)

4.4.2 VibViz Interface

Designed based on these requirements, *VibViz* is an interactive visualization with three views (Physical, Sensory/Emotional and Metaphor/Usage Example—Table 4.2), each with a screen area containing vibration representations and filter controls (Fig. 6.2a). Several features bear notice:

Linked views: All views show the same vibration subset at any time: a filter applied to one controls the others, and hovering over a vibration in one highlights that vibration elsewhere. Hovering over a tag in the tagclouds highlights associated vibrations in all three views.

Thumbnail design: A vibration glyph automatically highlights central characteristics of each vibration waveform and renders it as a thumbnail. The glyph encodes vibration frequency with colour saturation and a darker stroke envelope to highlight vibration pattern over time.

Drill-down and marking: A left or right click on a vibration respectively opens a detail popup (Fig. 4.3b), or bookmarks it. Marked vibrations have a highlighted border.

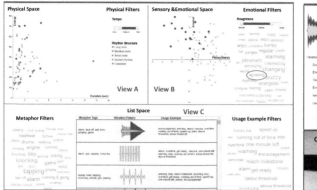
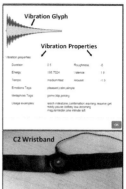

(a) *VibViz* interface- Hovering over a tag in any of the tagclouds (here, the "agitating" tag, circled in red, highlights the associated vibrations on all three views. This is done with: more saturated colors in view A, B and with a dark frame in view C. The labels "View A, B, C" are included for explanation and were not visible to participants.

(b)   Detailed   vibration popup in *VibViz* (top) and C2 wristband (bottom).

Fig. 4.3   The *VibViz* interface and C2 wristband that renders the vibrations

VibViz is best displayed on screen sizes equal to or larger than 12 inches and is designed for a single actuator. For multiple actuators, the user can playback one vibration simultaneously on several actuators or rely on the target application program to synchronize timings of the vibration notifications on multiple actuators.

4.4.3   Dataset

To use our vibration library in *VibViz*, each vibration had to be annotated for all five facets. We measured vibration duration, energy, and pulse structure. Three researchers annotated the other vibration properties; one annotated all and two half of the library. We averaged ratings and removed any pairs of contradicting tags.

4.5   User Study

We ran a small user study to investigate two questions:

(Q1) Does *VibViz satisfy its design requirements?* (Research tool; supports novice use, search, exploration, finding similar/contrasting items, serendipity, multiple pathways).

(Q2) How *useful* is each facet for personalization? How *interesting* is each for end-users? As pathways to exploring the library, does their *multiplicity* provide significant utility and interest over a single view?

Participants and Procedure: We recruited 12 participants (7 female) using flyers and social media posts, for a 1 h study and $10. The majority (8 out of 12) of the participants did not have any prior vibrotactile background beyond their cellphone vibration notifications. Three participants had attended vibrotactile demos or user studies in the past and one had experience in designing vibration patterns. We audio-recorded sessions and asked participants to verbalize their thoughts throughout.

In a pre-questionnaire, participants wrote down 1–2 daily activities and their preferred notifications (e.g., activity: running; notification: start and end of each interval). They then explored *VibViz* (displayed on a 14 inch laptop screen) for 10 min to get a sense of its features, while wearing a C2 tactor held in a wristband (Fig. 6.2a–c); the experimenter answered any questions about the interface. Participants next completed 9 scenarios (one at a time, 4 warm-up and 5 complex—Table 4.3), with random ordering in each set (≤3 min per scenario). Warm-up scenarios were clearly linked to one facet; complex scenarios were open-ended but common tasks in personalizing real world vibrotactile notifications and thus, were subject to interpretation. For example, the like/dislike scenarios were included to mimic situations where users' knowledge of the desired vibrotactile notification is purely implicit and visceral. Finally, participants filled a post-questionnaire. Throughout the session, the experimenter sat beside the participant and used an observation sheet to record confusions, comments, and actions taken to complete each scenario.

Data and Analysis: Our data consisted of demographics and notification types from pre-questionnaire, the experimenter's notes on confusions and list of actions for each scenario, and ratings and comments from the post-questionnaire. During the study, we noticed that sometimes participants used the *List*, *Physical*, or *Sensory/Emotional* spaces to explore the vibrations without using the characteristics of that facet. Thus, we analyzed participants' actions on filters and spaces separately.

Table 4.3 Study scenarios. Green/warm-up; blue/complex

Scenario	Description
Sc (Physical)	Find a vibration that is "short" in duration, "strong", and "fast".
Sc (Emotional)	Find a vibration that is "urgent" and "pleasant".
Sc (Metaphor)	Find a vibration that feels like a "fly or bee".
Sc (Usage Example)	Find a vibration that is good for both "start" and "stop" notifications.
Sc (Like)	Find a vibration that you like.
Sc (Not like)	Find a vibration that you do not like.
Sc (Pre-Q)	Find a vibration for the notification you wrote on the pre-questionnaire.
Sc (Combined)	Find a vibration that feels "natural", catches your attention, and is good for "every 5 minute notification".
Sc (Similar)	Find a vibration similar to the last vibration you chose.

Due to the study's small size and interesting variations among participants, we rely on summary statistics such as counts and percentages instead of statistical tests.

4.6 Results

We structure this section according to our research questions.

4.6.1 (Q1) Does VibViz Satisfy Our Design Requirements?

1-Serve as a research tool for vibrotactile researchers: *VibViz* provided adequate separation to allow us to observe and log participants' actions by facets. With the current design, however, one would need a combination of software logging and eye-tracking to automatically collect meaningful data.

2-Support novice users: Participant comments indicated that several terms and controls were confusing during initial exploration: *Rhythm structure* (10 participants), *Arousal* dimension (5), *AND/OR* filter operation (4). Also, none of the participants discovered the ability to bookmark vibrations or perform a zoom on the *Physical* view until they were told. 4 and 3 people respectively did not notice linked filtering or linked highlighting of vibrations across all views.

3-Support end-users in search and exploration tasks: According to post-questionnaire data, 9 participants followed "an explicit search" and 9 "a less-focused exploration" strategy, "many times" or "always", to find the vibrations. e.g., P1 stated that *"finding vibrations always started with an explicit search up to the point that I filtered everything that I thought might not be the proper ones for the scenario. Then I explored among the available filtered options"*.

4-Support users in finding similar vibrations: 6 participants used the visual vibrotactile glyphs and *List* space, 4 used proximity on the *Sensory/Emotional* space and 2 used *Metaphor* or *Usage Example* tags to find similar vibrations.

5-Facilitate serendipitous discoveries: Based on the definition of serendipity in [15], the frequency of finding a vibration "by accident" or "by a less-focused exploration" can be a measure of serendipitous discoveries. 8 participants found an interesting vibration "by accident", 9 found the scenario vibrations "by accident", and 11 found them "by a less-focused exploration" for at least "a few times".

6-Provide multiple pathways to the vibrotactile library: Based on the percentage of actions (Fig. 4.4), 7 participants used elements of at least two separate facets in more than 20% of actions. Participants also varied in their preferred filter and space combinations; e.g., P4 never used the *List* space, while P9 used it frequently (62%). All participants used different pathways for different tasks (Fig. 4.5). In our observations, these percentages also reflected the time the participants spent on the different parts of the interface.

	Physical	Emotional	Metaphor	Usage Example	Physical	Emotional	List
P1	0.24	0.16	0.13	0.04	0.16	0.05	0.23
P2	0.11	0.26	0.07	0.19	0.06	0.27	0.06
P3	0.09	0.22	0.06	0.07	0.00	0.00	0.56
P4	0.19	0.23	0.09	0.12	0.09	0.28	0.00
P5	0.19	0.26	0.07	0.22	0.12	0.1	0.04
P6	0.16	0.26	0.11	0.07	0.04	0.21	0.14
P7	0.19	0.02	0.27	0.06	0.06	0.00	0.38
P8	0.17	0.25	0.09	0.03	0.08	0.1	0.27
P9	0.11	0.11	0.06	0.00	0.02	0.1	0.62
P10	0.11	0.32	0.00	0.00	0.13	0.3	0.13
P11	0.04	0.27	0.11	0.07	0.22	0.13	0.15
P12	0.23	0.32	0.13	0.02	0.06	0.15	0.09

Fig. 4.4 Average filter and space usage per participant. Tan, yellow, and green colors denote low ($< 10\%$), medium ($< 20\%$), and high ($> 20\%$) usage frequency

All users	Physical	Emotional	Metaphor	Usage Example	Physical	Emotional	List
Sc(Physical)	0.57	0.09	0.02	0.02	0.21	0.07	0.04
Sc(Emotional)	0.17	0.41	0.02	0.08	0.04	0.22	0.07
Sc(Metaphor)	0.05	0.13	0.38	0	0.08	0.09	0.26
Sc(Usage Example)	0.1	0.16	0.07	0.33	0.08	0.13	0.14
Sc(Like)	0.05	0.35	0.06	0.02	0.03	0.2	0.3
Sc(Not Like)	0.05	0.3	0.02	0.05	0.03	0.23	0.33
Sc(Pre-Q)	0.19	0.13	0.08	0.07	0.18	0.05	0.3
Sc(Combined)	0.07	0.32	0.1	0.08	0.11	0.14	0.16
Sc(Similar)	0.12	0.13	0.14	0.03	0.03	0.15	0.39

Fig. 4.5 Average filter and space usage per scenario. Tan, yellow, and green colors denote low ($< 10\%$), medium ($< 20\%$), and high ($> 20\%$) usage frequency

4.6.2 (Q2) How Useful and Interesting Is Each Vibration Facet?

Facets interest and utility: According to post-questionnaire data (Fig. 4.6), the participants found the combination of all views most interesting, followed by *Sensory/Emotional*. *Physical* and *Usage Example* were least interesting. Similarly, all the views were perceived as useful, led by the full combination and *Sensory/Emotional*.

Frequency of facets use: In response to the question "Which of the following views would you use most often?", 8/12 participants chose *Sensory/Emotional*, 3 of whom wanted it in combination with the *Metaphor* or *Physical* views. According to P6, *"they are all useful for different things...I think I can use the Metaphor and Emotional view most of the time and occasionally switch to the other ones for a specific task"*. P8 had a similar comment. Among others, 3 selected the *Usage Example* and 1 the *Physical* view. Our observation data generally aligned with post-questionnaire data. On average, *Sensory/Emotional* filters were used most (22%), followed by *Physical* (15%), *Metaphor* (9%) and *Usage Example* filters (8%).

Mismatches: Post-questionnaire responses from P2, P7, and P9 conflicted with our observations. P2 chose *Usage Example* on the post-questionnaire but used *Sensory/Emotional* most often (26%). This difference was likely due to her stated dislike

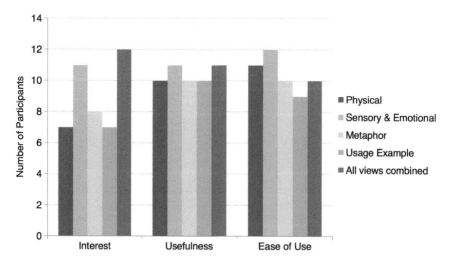

Fig. 4.6 Interest, usefulness, and ease of use for the vibrotactile facets based on the post-questionnaire data

for the tagcloud design for the *Usage Example* filters. P9 chose *Sensory/Emotional* but mostly used the *List* space (62%) during the scenarios, noting that *"I want to go through them all, don't wanna miss some by filtering."* Most curiously, P7 chose *Usage Example* but used it the least during the study. We cannot speculate on the reason. We did not notice any differences in the usage patterns of the four participants who had attended vibrotactile demos or user studies or had vibrotactile design experience.

Other useful features: Visual vibrotactile glyphs were appreciated (9/12 rated them as somewhat or very useful). In our observation, they were especially helpful for finding a previously seen/felt vibration, and for finding similar vibrations. According to P4, *"Based on the visual pattern, I started to realize which ones I like and don't like."* The *List* space was also used frequently (22%) for going through all the remaining vibrations. Also, P3, and P9 mainly used the *List* space for the complex scenarios since they felt that their perception of vibrations did not match some of the tags.

4.7 Discussion

4.7.1 Interface Requirements

Our study results suggest **several features that are important for a vibrotactile library navigation**: (1) filtering functionality, (2) visual vibration pattern, (3) spatial and tabular presentations, (4) bookmarking, and (5) simple vibrotactile authoring tools.

We found that filters supported the search task and helped users narrow down to a vibrotactile subset that matched their criteria, while the visual vibration glyphs, list (tabular), and spatial representations were most useful for exploration. The spatial and tabular representations allowed the users to flexibly sample the library, but the visual vibration glyphs made this exploration quicker and also assisted in similarity search. In some cases, participants wanted to adjust the sensation of a vibration; this calls for **incorporating simple authoring tools** into vibrotactile library navigation interfaces (Chap. 3).

4.7.2 Vibrotactile Facets

Keep all, show a subset, allow switching: Although the majority of users found a combination of facets most interesting and useful and used all the facets at some point, most often each only used about two views. Thus, we think the library navigation interface could show a subset of views to the users but allow them to switch to other views as needed. Reducing the number of views frees up screen space for other useful functionality (e.g., a personal view for a favorite vibration subset or for temporary comparison) and makes the tool viable for smaller screen sizes.

Support personalization: Users appear to vary in which subset of the views they prefer. Thus, supporting personalization of default views is an important requirement. If only a single facets can be incorporated, our results suggest that the *Sensory/Emotional* view is a reasonable default.

4.8 Conclusions and Future Work

We developed and studied five organization and navigation schemes (vibrotactile facets) for a library of 120 vibrations. We designed *VibViz*, an interactive library navigation tool, to: (1) support novice end-users in personalizing vibrotactile notifications, and (2) serve us as a research tool for studying the utility and appeal of the facets. Our user study with 12 participants found greatest interest in the *Sensory/Emotional* facets, but also interesting variations among participants in preference for all the facets. Our results revealed the importance of visual scanning (tabular and spatial overview, and visual vibrotactile pattern) for efficient library navigation.

In the next chapter, we report our process for collecting library annotations from a large group of users and analyze variations in their ratings and usage. Future tools can extend *VibViz* to support additional personalization tasks, such as vibration set creation and item comparisons.

References

1. Brewster, S., Brown, L.M.: Tactons: structured tactile messages for non-visual information display. In: Proceedings of the Fifth Conference on Australasian User Interface, vol. 28, pp. 15–23. Australian Computer Society, Inc. (2004)
2. MacLean, K., Enriquez, M.: Perceptual design of haptic icons. In: Proceedings of EuroHaptics Conference, pp. 351–363 (2003)
3. Tam, D., MacLean, K.E., McGrenere, J., Kuchenbecker, K.J.: The design and field observation of a haptic notification system for timing awareness during oral presentations. In: Proceedings of the ACM SIGCHI Conference on Human Factors in Computing Systems (CHI '13), pp. 1689–1698. ACM, New York, NY, USA (2013). https://doi.org/10.1145/2470654.2466223
4. Schneider, O., Zhao, S., Israr, A.: Feelcraft: User-crafted tactile content. In: Haptic Interaction, pp. 253–259. Springer (2015)
5. Israr, A., Zhao, S., Schwalje, K., Klatzky, R., Lehman, J.: Feel effects: enriching storytelling with haptic feedback. ACM Trans. Appl. Percept. (TAP) **11**, 11:1–11:17 (2014)
6. Immersion corporation: Haptic effect preview, http://www2.immersion.com/developers/. https://play.google.com/store/apps/details?id=com.immersion.EffectPreview&hl=en. Accessed 24 Jan 2015
7. Immersion corporation: Haptic muse, http://www2.immersion.com/developers/. http://ir.immersion.com/releasedetail.cfm?ReleaseID=776428. Accessed 24 Jan 2015
8. van Erp, J.B., Spapé, M.M.: Distilling the underlying dimensions of tactile melodies. In: Proceedings of Eurohaptics Conference, vol. 2003, pp. 111–120 (2003)
9. Ternes, D., Maclean, K.E.: Designing large sets of haptic icons with rhythm. In: Haptics: Perception, Devices and Scenarios, pp. 199–208. Springer (2008)
10. MacLean, K.E.: Foundations of transparency in tactile information design. IEEE Trans. Haptics (ToH) **1**(2), 84–95 (2008)
11. Obrist, M., Seah, S.A., Subramanian, S.: Talking about tactile experiences. In: Proceedings of the ACM SIGCHI Conference on Human Factors in Computing Systems (CHI '13), pp. 1659–1668. ACM (2013)
12. Schneider, O.S., MacLean, K.E.: Improvising design with a haptic instrument. In: Proceedings of IEEE Haptics Symposium (HAPTICS '14), pp. 327–332. IEEE (2014)
13. Guest, S., Dessirier, J.M., Mehrabyan, A., McGlone, F., Essick, G., Gescheider, G., Fontana, A., Xiong, R., Ackerley, R., Blot, K.: The development and validation of sensory and emotional scales of touch perception. Atten., Percept., Psychophys. **73**(2), 531–550 (2011)
14. Schneider, O., MacLean, K.E.: Haptic jazz: collaborative touch with the haptic instrument. In: Proceedings of the IEEE Haptics Symposium (HAPTICS '14). IEEE (2014)
15. Thudt, A., Hinrichs, U., Carpendale, S.: The bohemian bookshelf: supporting serendipitous book discoveries through information visualization. In: Proceedings of the ACM SIGCHI Conference on Human Factors in Computing Systems (CHI '12), pp. 1461–1470. ACM (2012)
16. Musicovery: Musicovery. http://musicovery.com/. Accessed 21 Oct 2016
17. Munzner, T.: Visualization Analysis and Design. CRC Press, Boca Raton (2014)
18. Engineering Acoustics, Inc.: C2 tactor, https://www.eaiinfo.com/. https://www.eaiinfo.com/tactor-info/. Accessed 21 Mar 2017
19. The Noun Project, Inc.: Nounproject, http://thenounproject.com/. Accessed 24 Jan 2015
20. Mazzoni, D., Dannenberg, R.: Audacity software, http://audacity.sourceforge.net/. Accessed 24 Jan 2015

Chapter 5
Deriving Semantics and Interlinkages of Facets

Abstract *Haptic facets* (categories of attributes that characterize collection items in different ways) are a way to describe, navigate and analyze the cognitive frameworks by which users make sense of qualitative and affective characteristics of haptic sensations. Embedded in tools, they will provide designers and end-users interested in customization with a road-mapped perceptual and cognitive design space. In the previous chapter, we compiled five haptic facets based on how people describe vibrations: *physical*, *sensory*, *emotional*, *metaphoric*, and *usage examples*. Here, we report a study in which we deployed these facets to identify underlying dimensions and cross-linkages in participants' perception of a 120-item vibration library. We found that the facets are crosslinked in people's minds, and discuss three scenarios where the facet-based organizational schemes, their linkages and consequent redundancies can support design, evaluation and personalization of expressive vibrotactile effects. Furthermore, we report between-subject variation (individual differences) and within-subject consistency (reliability) in participants' rating and tagging patterns to inform future progress on haptic evaluation. This facet-based approach is also applicable to other kinds of haptic sensations. Finally, we detail our novel methodology for collecting user annotations for a large haptic collection in the lab.

5.1 Introduction

Despite growing interest in and availability of haptic technology in consumer markets, even its most common manifestation of vibrotactile feedback is still limited in everyday use, generally appearing as a dull, undifferentiated and often annoying buzz. While a dearth of expressive hardware is one obvious cause, there are comparable difficulties in *designing* with even the hardware we already have for both vibrotactile and other haptic display modalities [1].

Design is difficult for many reasons, not least due to large variances in individuals' preference and interpretation of how vibrations feel and what they suggest [2–5]. Here we highlight two gaps in support which we propose are central.

© Springer Nature Switzerland AG 2019

H. Seifi, *Personalizing Haptics*, Springer Series on Touch and Haptic Systems, https://doi.org/10.1007/978-3-030-11379-7_5

Design- How to make a "surprise" vibration?

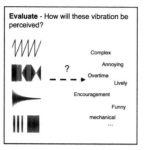

Evaluate - How will these vibration be perceived?

Complex
Annoying
Overtime
Lively
Encouragement
Funny
mechanical
...

Personalize – I want to make my message notifications "lively".

(a) **Design Guidelines and Refining**: Designers often need to translate aesthetic requirements specified in emotion, metaphor, and usage spaces (*e.g.,* **surprise**) to sensory and engineering parameters (*e.g.,* frequency); and to refine candidates.

(b) **Evaluation:** Assessing or accessing the perceptual and aesthetic qualities of vibrations, created by manipulating engineering parameters, allows designers to use them appropriately.

(c) **Personalization:** End-users can more efficiently select and tune vibrations in a perceptual and aesthetic space than in an engineering space, requiring the further capability of *repositioning* sensations within cognitive spaces.

Fig. 5.1 Three scenarios in vibrotactile design, evaluation, and personalization that facets can support when fully instantiated in design tools

A Lack of Guidelines and Tools: When making (sketching, refining) and evaluating sensations, designers often identify requirements in terms of usage examples (e.g., allowing presenters to track time during their presentations), intended emotions (sadness, surprise), or accompanying media (a racing car in a game) [6–10], but are forced to *design* with engineering parameters (Scenario 1, Fig. 5.1a). In other cases, designers have a set of vibrations (whether newly created or accessed within an existing collection) and wish to *evaluate* their aesthetic and qualitative characteristics (Scenario 2, Fig. 5.1b). The ability to use low-level engineering parameters to construct or evaluate for affective and qualitative characteristics is tacit knowledge that haptics designers build over years and through extensive contact with users. It is hard to communicate, incorporate in tools or transfer to others.

Perception is Personal but Personalization is Unsupported: Past studies of vibrotactile applications in real-world contexts indicate the necessity of end-user personalization [7, 10]. However, there is a dearth even of effective *expert* tools for far more accessible and perceptually understood engineering parameters like vibrotactile amplitude and frequency; easy and practical mechanisms that would make sense to end-users are rare indeed. Unsurprisingly, previous work suggests that personalizing based on engineering parameters is beyond end-user capacity and willingness. When given tools in their own language domain, users can quickly access and modify their desired vibrotactile notifications (Scenario 3, Fig. 5.1c, and Chaps. 3 and 4).

5.1.1 Facets: Aligning Content Access with Mental Frameworks

People unconsciously use a multiplicity of cognitive frameworks or *schemas* to describe qualitative and aesthetic attributes of vibrations (Fig. 5.2) [11, 12]. Sometimes people describe a vibration based on its similarity to something they have experienced before (*this is like a cat purring*), on emotions and feelings (*this is boring*), or intended usage (*this tells me to speed up*). These schemas, themselves composed of many attributes (Fig. 5.3a) are in users' minds: shaped by their past experiences and training, they provide a cognitive scaffolding on which people rely for sense-making.

Facets, a design concept originating from the information retrieval domain [13–17], capture the multiplicity and flexibility of users' sense-making schemas for physical and virtual items. A facet encapsulates the properties or labels related to one aspect of or perspective on an item and offers a categorization mechanism. For example, examples of alternative facets for a collection of architectural images are people (such as designer, agency, historical figure), time periods, geographical location (GPS coordinates, province, neighborhood), and structure types (function, architectural elements). For a collection of clothing items they might be garment type (top, bottom, inner, outer, accessories), color, brand, formality, season [13, 15]. A given facet may be composed of a single property (e.g., brand) or a set of diverse elements that reflect that perspective—e.g., lists of descriptive words (tags), numerical scales, binary or multicategory attributes (e.g., province). The facet characterization varies by domain and relies on a user's knowledge and conceptual mapping of that domain.

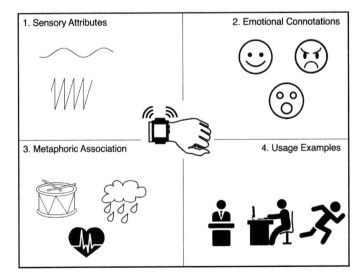

Fig. 5.2 Conceptual sketch of the four vibration facets

Multiple facets can be used flexibly together to describe or examine different aspects of a given item in a collection, or alternatively, explore those aspects in light of other collection items.

In Chap. 4, we identified five facets for vibrations based on the literature which captured: (1) *physical* attributes of vibrations that can be objectively measured such as duration, rhythm structure, etc. (2) *sensory* properties such as roughness, (3) *emotional* connotations, (4) *metaphors* that relate the vibration's feel to familiar examples, and (5) *usage examples* or events where a vibration fits (e.g., "speed up"). We implemented these facets in an interactive graphical visualization and navigation tool, *VibViz* (Chap. 4).

Here, we revise these into four facets: sensation, emotion, metaphor, and usage examples (Table 5.1). For consistency with past haptic literature [18], we now refer to dimensional attributes that can be objectively measured (e.g., duration, frequency) as *engineering space*. The sensation facet now includes the subjective dimensional attributes energy and tempo, previously in the physical facet.

These facets provide unique ways to assign a familiar meaning to a haptic sensation. For example, the metaphor and usage facets rely on previously experienced

Table 5.1 Vibration facets used here, taken with minor alterations (†) from Chap. 4. These facet properties are combinations of ratings (quantitative attributes such as i,ii, iii for sensation facet) and tags (list of words iv). For example, in the sensation facet, *i*, *ii* and *iii* are single attributes on which an item can be rated, while *iv* is a list of descriptive tag words that might apply to sensations when considered from this viewpoint. *Modifications:* (1) Omitted the physical facet. For consistency with past haptic literature [20], we now refer to dimensional attributes that can be objectively measured (e.g., duration, frequency) as *engineering space*. (2) The sensation facet now includes the subjective dimensional attributes energy and tempo, previously in the physical facet.

Facet	Attributes
1. Sensation Perceptual properties of vibration	(i) *Energy*†
	(ii) *Tempo* or speed†
	(iii) *Roughness*
	(iv) *Sensory words*: *24 adjectives* from touch dictionary [21]
2. Emotion Emotional interpretations of vibration	(i) *Pleasantness*
	(ii) *Arousal*
	(iii) *Emotion words*: *26 adjectives* from touch dictionary [21]
3. Metaphor Familiar examples resembling the vibration's feel	*Metaphor words*: We collected a set of *45 metaphors* for our list of usage examples, asked colleagues and friends to provide metaphors for our vibrotactile effects, and used the NounProject website [22] for brainstorming on metaphors
4. Usage Examples Types of events which a vibration fits	*Usage example words*: We collected and consolidated a set of *24 usage examples* for presentation timing and exercise tracking [7]

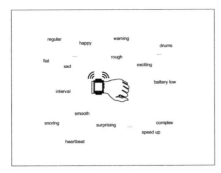

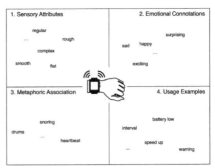

(a) People use **mixed language** to describe and make sense of vibrations, which is highly descriptive; but its disorganization makes it hard to use in design.

(b) **Facets** organize users' descriptions into *categories of labels*, each describing and orienting elements according to one aspect that the labels in that facet share.

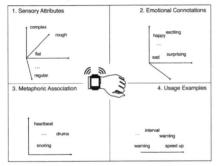

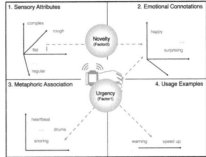

(c) The underlying **semantic dimensions** of each facet (shown as black arrows) structures its attributes, and exposes axes along which there is continuity.

(d) Factors $_{fact}$ are conceptual constructs that can describe the **linkages between dimensions of the four facets** (red arrows)).

Fig. 5.3 Concept sketch showing haptic facets, dimensions and their linkages. Central elements (denoted throughout the chapter with a special font and subscripts) include (1) tag: a label/word that people use to describe an attribute of a haptic sensation (e.g., soft, exciting); (2) facet $_f$: a framework that binds related attributes of haptic sensations into a descriptive category; (3) dimension $_d$: a continuous parameter that delineates variations in a facet; and (4) factor $_{fact}$: a conceptual construct underlying linkages among different facets (deduced here using factor analysis)

sensations and usage contexts to make sense of vibrations (see [19] for more details). We implemented these facets in an interactive graphical visualization and navigation tool, *VibViz* [19], and denote them and related concepts here with a special font and subscripts (as explained in Fig. 5.3).

While not meant to be a unique or complete delineation of the possible vibrotactile facet space, this set does provide a practical sense of what facets can offer to design. Because a given vibration can be located in the context of any and all, each highlighting a particular aspect, they can organize a messy hodgepodge of inconsistent language and mixed models into a powerful tool that leverages perception and analogy. The interactive visualization tool *VibViz* allows untrained users to peruse a large vibrotactile collection by viewing items in multiple facets simultaneously and dynamically.

These multi-facet views thereby become rich, layered descriptions which inform design. For example, *VibViz*'s linked facets show how an individual item may have different perceptual near-neighbors and contrasts in the different facets.

From Browsing to Manipulating in Facet Space: In its primary form, a facet is just a flat list of attributes like tags and ratings (Fig. 5.3b). Thus, it only allows us to browse existing, defined elements (as *VibViz* does). What if a designer or user wants to *change* an element, or find points in between existing library items (Fig. 5.1 scenarios)? A semantic *dimension* offers a structure for the facet; it provides a continuous perceptual parameter along which one can move vibrations or characterize them (Fig. 5.3c). Imagine a slider that makes a vibration more or less "exciting", "alluring" or "bell-like"—in contrast to ones that change its base frequency or amplitude. Such sliders would allow both trained designers and untrained end-users to manipulate (sketch, ideate, personalize) vibrotactile signals more directly by offering handles in a language framework relevant to their purpose.

However, to allow continuous movement along cognitively useful dimensions, a tool must do far more than locate discrete sensations within facet space: *it must identify and present a topologically continuous mapping between the facets and engineering spaces, so that every point of the slider's range can be rendered.*

Further, *VibViz* already hints at considerable redundancy between facets—when a dimension in one facet is very similar to that of another, but goes by a different name. Facets are not independent spaces, but alternative views of the same thing. Mapping connections specifically will enable designers to translate or formulate requirements from one facet space (e.g., emotional or application-driven constraints) into more actionable sensory and engineering spaces (*Scenario 1*, Fig. 5.1a) or evaluate aesthetic characteristics of a set of vibrations given their sensory properties (*Scenario 2*, Fig. 5.1b).

5.1.2 Research Questions

A major objective of this research is to establish a means of finding such mappings. As a first step, we have pursued three questions:

(Q1) Within-Facet Substructure: *What are the underlying dimensions of the facets that dominate users' reaction to vibrations?* For example, for the emotion facet one could then design or identify the most emotionally distinct vibrations. These dimensions are the first step towards perceptually salient continuous "sliders", such as roughness.

(Q2) Between-Facet Linkages: *How are attributes and dimensions in different facets linked with each other?* A specific mapping will allow for translation of requirements from one facet to another (e.g., emotion to sensation and vice versa) and provide the basis for a topologically continuous mapping between the facet dimensions and engineering parameters. Designing a "surprising" sensation is much simpler if one can access its sensory characteristics to be irregular, ramping up, and rough. Our format convention for vibration tags or attributes highlights *points* in a facet space, as opposed to dimensions.

(Q3) Individual differences in facets: *To what extent do people coincide or differ in their assessment of vibration attributes?* Facets are based on frameworks in users' mind which can vary greatly, for example due to past experiences and culture. Understanding this variation can shed light on individual differences in preferences and meaning-mappings, and inform development of robust haptic evaluation instruments.

5.1.3 Scope

We used the *VibViz* vibration library and the concept of facets to investigate these questions. We first collected an extensive set of user *annotations* (selections of adjective ratings and tags) for library elements to situate the vibrations within the four facets (Chap. 4). We obtained this data in a two-step process adapted from data collection methods in the music domain [23], first with three experts and then with 44 lay users.

In our subsequent analysis, we derived semantic dimensions of each facet through Multidimensional Scaling (MDS) analysis [24], and investigated between-facet linkages using factor analysis [25]. With this data, we updated and further populated Table 5.1's descriptions to include our derived facet dimensions and their linkages. Our analysis occurred at multiple levels: we examined low-level properties and linkages of individual tags (*tag level*), and then semantic facet dimensions obtained from MDS analysis (*dimensional level*), and finally compared these across the four facets (*facet level*). Thus, our novel contributions include:

1. Empirically derived semantic dimensions of four vibrotactile facets;
2. Between-facet linkages at dimensional and individual tag levels, and discussion of their implications for vibrotactile design and tools;
3. Analysis of individual variations in rating and annotating vibrations;
4. A two-step methodology for annotating large sets of vibrotactile effects, and data on its validity and reliability; and
5. A publicly available dataset of 120 vibrations and their annotations and dimensions [26].

In the remainder of the chapter, we present the related literature on tool development, perceptual dimensions of vibrations, and haptic evaluation methodology (Sect. 5.2), and highlight important aspects of our approach (Sect. 5.3) followed by data collection details (Sect. 5.4) and analysis procedure and results (Sect. 5.5). In Sect. 5.6, we describe how our results support the design and evaluation scenarios outlined above

(Fig. 5.1) and compare our facet dimensions and linkages to any existing dimensions in the literature. We finish by reviewing our data collection and analysis methodology and presenting interesting directions for future work.

5.2   Related Work

The design process for haptic sensations will inevitably vary substantially depending on designers and use cases, but it usually involves several rounds of design, evaluation, and fine tuning of the stimuli and usage scenarios [1, 6, 27, 28]. To support this process better, we need effective authoring tools, design knowledge and guidelines, as well as evaluation methodology and metrics. Below we describe progress in these areas and how our work builds on them.

5.2.1   Tools for Vibrotactile Design and Personalization

With their crucial role in the design process, haptic authoring tools have received an increasing attention in the last decade. Design tools by nature facilitate use of some parameters and approaches while limiting access to others; e.g., pre-designed themed color sets versus full-spectrum palettes—an example of parameter-limiting; or fine tuning and precision versus rapid prototyping and creative flow, i.e., approach-limiting. Existing haptic tools are built around the most important design parameters and approaches identified in the literature or by practitioners. For example, to support design around rhythm or temporal pattern, the tools facilitate precise modification and referencing of vibrations on a timeline [9, 29, 30]. Recent instances promote use of examples and design by demonstration as well as rapid prototyping by allowing easy modification of design parameters [30–32]. However, to our knowledge these tools currently provide access only to low-level engineering parameters. Perceptual and affective controls over vibrations are missing, and this slows design.

Content design and manipulation are no longer done only by a specific group of users [33]. In several other domains (e.g., photo and video editing, music mixing, configuring software), a spectrum of tools exist for various expertise levels [34–36]. Haptic design tools are catching up: while past tools have mostly focused on experts, recent trends, published during this PhD work, have targeted end-user haptic content creation and personalization [8, 10].

Our work informs design of higher level controls, which can be thought of as tuning sliders or knobs and might be implemented as such in a design interface. These will benefit both expert design tools and end-user personalization interfaces.

5.2.2 Knowledge of Perceptual and Qualitative Attributes of Vibrations

A body of work has investigated perceptual dimensions of natural (e.g., textures) and computer-rendered synthetic haptic stimuli (e.g., vibrations), and users' language for touch [3, 11, 21, 37–40]. In our own previous work, *VibViz*, we compiled five vibrotactile facets based on dimensions and properties known in the literature for vibrations and users' language (Chap. 4).

Several tactile perceptual studies exist on natural textures (e.g., fabrics, fluids and various surface materials) due to their higher availability and wider range of sensations (see [39] for a survey). However, the resulting dimensions (such as warm/cold) are not easily translated to computer-rendered synthetic sensations. Others examine prominent vibrotactile attributes based on users' similarity groupings or ratings for small to large sets of vibrations. They report energy, roughness and rhythm as the most important design parameters [18, 37, 41, 42]. While these studies give insights into vibration perception, they tend to be organized in terms of engineering or sensory parameters and are not linked to language attributes in users' minds.

Recent studies examine users' tactile language and descriptions as a window onto understanding prominent properties of touch. Notable among these is Guest et al.'s collection of touch-related English vocabulary [21]: based on MDS analysis of word similarities, the authors propose three dimensions for sensory words (*roughness*, *dryness* and *warmness*), and three for emotional words (*comfort*, *arousal* and *sensual quality*). We use this collection of sensation and emotion words in our facets; however, the identified dimensions are not validated for synthetic haptic sensations. Further, other aspects of users' languages such as metaphors and usage examples are not examined.

Our own facets were previously constructed based in part on this literature; here, we further confirm, refine and add to these dimensions and attributes by analyzing users' perception of a large library of vibrations collected through the facets.

5.2.3 Methodology for Evaluating Qualitative Attributes of Vibrations

Previous research in related areas typically adapts methodology from other domains for haptic studies, or refines existing haptic evaluation methodology to be more time- and cost-effective. For example, MDS studies in haptics were originally adapted from the auditory domain to investigate perceptual distances between tactile sensations [3, 24, 43]. Other researchers use phenomenology to obtain richer language-based descriptions of haptic sensations [11, 12]. However, phenomenological studies are time-consuming and thus are only practical with few participants and small sets of sensations. In Chap. 6, we examine the feasibility of using crowdsourcing platforms (e.g., Amazon's Mechanical Turk) for vibration evaluation. Despite promising results, the methodology is mainly tested for Likert scale evaluation and is not yet verified for richer, language and annotation-based haptic studies.

Despite some progress in haptic evaluation approaches, it remains singularly difficult for a researcher to collect rich feedback from lay users in a manner that scales to large stimuli sets. Our data collection methodology, adapted from the music domain, by necessity has had to fill this gap. Here, we report its execution details and examine its validity and reliability.

5.2.4 Instruments for Evaluating Haptic Sensations

As haptic effects are designed for a wide variety of use cases and requirements, researchers frequently must devise a custom evaluation instrument for every study. Recent investigations have laid the foundations for devising a standard yet flexible instrument for vibrations through examining users' language and compiling important vibration properties and common metrics across past studies.

Most relevantly, Guest et al. provide a linguistic instrument for tactile sensations called the "touch perception task" (TPT) [21]. TPT is composed of 26 sensory ratings and 14 emotional ratings and was tested by its authors on natural textures.

Here, we have re-used the annotation instrument we previously developed for validating and populating *VibViz*, built around language and metrics found in the literature. Specifically, (a) four of our five Likert scale ratings (strength/energy$_d$, roughness$_d$, pleasantness$_d$, and arousal$_d$) are commonly used metrics; while (b) our sensation$_f$ and emotion$_f$ tag lists are based on Guest et al.'s sensation and emotion vocabulary [21]. We introduced the tempo$_d$ rating scale as well as the metaphor$_f$ and usage example$_f$ tag lists in the previous chapter on *VibViz* (Chap. 4). When used to annotate a large vibrotactile library, this more comprehensive annotation instrument can generate results that will inform future vibrotactile evaluation instruments by identifying the redundant facet attributes and providing an estimate of users' reliability and variation in responses.

5.3 Approach

To investigate the semantic dimensions of these facets and their linkages, we began with *VibViz*'s source vibrations and its comprehensive but efficient evaluation instrument (Chap. 4). We report the scalable and robust methodology that allowed a comprehensive annotation of our vibration library and use standard dimensionality reduction methods to analyze the resulting dataset. Below, we describe each aspect of our approach in more detail.

5.3.1 Rich Source Vibrations

To identify underlying dimensions and linkages of facets, we used a large and varied set of vibrations. In Chap. 4, we described our various tools and inspirations including systematically changing vibration parameters (e.g., rhythm, frequency), modifying audio files to serve as vibrations or using audio files as reference for designing vibrations, and running pilot design studies where our lab colleagues designed vibrations for a given set of metaphors (see Chap. 4 for more details on our library design process). Our design process was intertwined with developing the four facets and their annotation instrument and resulted in 120 vibrations with a wide range of qualitative and affective characteristics.

5.3.2 Inclusive and Concise Annotation Instrument, for a Flat Descriptor Set

For an accurate picture of the vibrations, we needed an inclusive and non-redundant annotation instrument. If an important rating or tag is not included, we would be unable to identify the corresponding dimension (exclusion risk). In contrast, redundant ratings or tags can introduce noise. As the set of ratings and tags grows, users' (even experts') ability to consistently characterize a vibration decreases (redundancy risk).

We developed our ratings and tags to reduce both risks. We included quantitative rating scales that are frequently utilized in the literature and incorporated as many relevant tags as possible in our evaluation's first step with experts (mitigating exclusion risk), and after the expert annotation phase, removed and consolidated redundant items in a discussion session (mitigating redundancy risk). The ratings capture users' perception on attributes that are previously identified to be salient for vibrations, while the tags allow us to derive salient dimensions not known before. The results of the process are five bipolar 7-point Likert scale ratings and four lists of candidate tags (see Table 5.1 for an overview, and Sect. 9.1 for a full list of tags proposed for each facet).

5.3.3 Scalable and Robust Data Collection Methodology

We needed a comprehensive 'gold standard' annotation set that covered the full *VibViz* library. Ideally, annotations would be applied by individuals who rated the entire facet space for all the items. This would require individuals rate and tag 120 vibrations, each according to five scales and 121 candidate tags. In piloting, we found this was too mentally and physically demanding to be suitable for lay users with varying levels of commitment, confirmed by poor signal-to-noise properties of that pilot data. We therefore devised a new collection method that could be spread across multiple participants (scalable) and would be robust to outliers, i.e., the occasional low-commitment participant—or at least, to clearly identify these.

Music annotation literature provides interesting alternative approaches for data collection, such as a *panel of experts*: Pandora Internet Radio uses experts to annotate its music dataset, constructing a "gene sequence" for each music piece that is used for music recommendations [23, 44]. Alternatively, services such as Last.fm *crowdsource* the annotation task, incenting end-users to add free-form textual tags to songs from which it derives music "folksonomies" [23, 45, 46]. However, our access to haptics experts is limited and the literature lacks a set of standard attributes for vibrations. Furthermore, we can not yet fully crowdsource vibration annotations, in large part due to hardware limitations and lack of a validated methodology (Chap. 6).

We therefore adapted these two approaches into a two-stage evaluation system. In the first *expert annotation* stage, three haptics designers rated and tagged the vibrations employing initial rating scales and tag lists, with encouragement to be liberal in application of tags to stimuli. In the *lay user validation* stage, a larger number of participants with no haptic background adjusted the experts' ratings and tags for subsets of the library—principally by removing tags which they felt did not apply, since this proved to be mentally easier than applying new ones; although tag addition was also allowed. The first stage resulted in consistent annotations across the library that were relatively free of the noise introduced by participants' fatigue and lack of commitment, but reflected only a small number of subjective opinions. In the second stage, we pruned the potentially overpopulated annotation dataset by bringing in additional, but potentially less committed, perspectives. We fully detail the process in Sect. 5.4.

This methodology does have a bias risk: participant perceptions of vibrations in the second stage can be influenced by the rating values and tag assignments that they are shown. We devised mechanisms in our experiment design to mitigate this bias, and evaluated its impact on our final dataset.

5.3.4   Data Analysis Methods

We used Multidimensional Scaling to identify the underlying dimensions for the tags (but not the rating scales or values) in each facet, and factor analysis to investigate constructs that link dimensions (including rating scale data) between various facets.

Multidimensional Scaling is a dimensionality reduction technique that is commonly used to derive and visualize a low-dimensional perceptual space from a high-dimensional dataset [24]. We used Matlab's classical MDS implementation (a.k.a. Principal Component Analysis or PCA) where the distances among the items (vibrations or tags) are Euclidean—as opposed to ordinal, as in a non-metric MDS [47].

Factor analysis is typically used to identify underlying variables (a.k.a. factors) that connect and describe a set of observed but correlated quantitative variables

[25, 48]. For example, factor analysis is usually applied to surveys with several Likert-scale questions to find connected questions. We applied factor analysis to our derived facet dimensions, and the ratings collected for our five scales.

5.4 Data Collection and Pre-processing

Here, we detail the collection of ratings and tags for the vibrations in two stages described above—expert annotation, and novice validation; then describe dataset pre-processing and define the metrics with which we analyzed its tags and ratings.

5.4.1 Stage 1: Annotation by Haptics Experts

We required expert annotators who had experience with a wide range of haptic and/or vibrotactile sensations, were familiar with our vibrotactile library and facets, and could commit to annotate all or a large subset of the vibrotactile library within a short time span of a few days. Within-subject annotation of the entire vibration set would produce consistency and breadth in our initial annotation dataset; however it did impose a substantial cognitive load on the expert annotators, and thus we utilized experts with some commitment to the research and group. Given the nature of the task, we did not feel this closeness to the research could bias the results, but leveraged it for motivation.

Expert backgrounds: Three haptic researchers including the first author provided expert annotations. The first author, a vibrotactile researcher who developed the vibration library and annotation instrument, rated and tagged all the vibrations while the second and third experts each annotated half of the vibrations (randomly assigned to them). The second annotator is a haptic researcher at University of British Columbia with extensive experience in designing and evaluating vibrotactile sensations and haptic design tools, The last annotator is a Human-Computer-Interaction researcher who co-designed *VibViz* with the authors and had extensive exposure to all the vibrations in the library before participating in this study. The second and third annotators received a $50 honorarium for their participation.

Initial dataset: 120 vibrations from *VibViz* library were randomly divided into 10 groups with 12 vibrations in each group. These groups remained fixed for all three expert annotators.

Apparatus and procedure: The annotation interface was a web-based wizard that gradually disclosed the available ratings and tags for the vibrations on subsequent tabs. The first tab disclosed five rating scales (7-point Likert scales) for the vibrations (Fig. 5.4b, Table 5.1). The four other tabs had the list of tags for the sensation$_f$, emotion$_f$, metaphor$_f$, and usage example$_f$ facets plus a textbox for any additional tags from the experts (Fig. 5.4b). In each session, first the experts played a fixed set of representative vibrations for calibration purposes, then proceeded to annotating

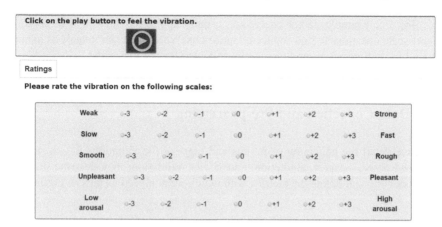

(a) The first tab shows all the five rating scales.

(b) The other four tabs show a list of potential tags in each facet and a textbox at the top for extra tags.

Fig. 5.4 Expert annotation interface- One can play a vibration many times and move between the tabs representing the required ratings and tags for the vibration, but they cannot go back to previous vibration(s)

a group of 12 vibrations (randomized presentation order). During the annotation process, the experts could play a vibration several times and move between different tabs for one vibration but they could not go back to previous vibration(s), even within that group. At the end of each group, a review page showed all the expert's ratings and tags for the vibrations which could be further edited. This procedure encouraged the experts to focus on the demanding task of annotating each vibration individually but also allowed for cross comparisons and consistency adjustments afterwards.

Annotating a group took about 45–60 min. Experts were given the choice of annotating their groups in a single session or spread over several sessions, but were not permitted to interrupt a single group's annotation. Expert 1, the first author, evaluated 10 groups over 5 sessions within 6 days, while Experts 2 and 3 evaluated their 5 groups over 3/8 and 4/4 sessions/days, respectively. The experts were allowed to revisit their previously annotated groups (but never did request to do so). The total time spent by each expert was approximately 8 h for Expert 1, and 4–5 h for Experts 2 and 3.

Pre-processing and tag consensus and consolidation: After collecting all the annotations, the first author examined all the tags for each vibration and highlighted conflicting tags (e.g., smooth tag by one expert and rough by another one, or angry vs. happy). In a follow-up session, all three experts played and felt vibrations with contradictory tags again and came to consensus on one of the conflicting tags or on removing both. Further, they could and did adjust wording (e.g., dynamic instead of changing), and combined tags under one wording (e.g., jaggy and grainy were replaced by grainy).

5.4.2 Stage 2: Validation of the Dataset by Lay Users

Our sole requirement for our Stage 2 participants was to have no background in haptics beyond normal everyday exposure to vibration sensations (e.g., via cellphone usage).

Participants and compensation: We recruited 44 participants (24 female, 19–60 years old, with 40 of the participants under 36 years old) through advertising on a North American university campus. All participants were university students except for three who did not declare their occupation. Participants were permitted to participate in more than one session but tag different vibrations in each session (up to a maximum of 4 sessions covering all 120 vibrations) and six participants did so. Participants were compensated $10 for a one-hour session.

Initial dataset: Our dataset was composed of the 120 vibrations with the average expert ratings and the combined and consolidated tags for each vibration, randomly divided into 12 groups of 10 vibrations. This grouping remained fixed for all the participants.

Mitigating bias and noise in the validation stage: We anticipated that the existing ratings and tags could bias participants' perception of the vibrations and/or suggest a lower need for their attention. Following literature guidelines on detecting invalid responses [49, 50], we mitigated this by making additions to the database which would *expose* non-diligent participants, and warned participants of the possibility of inconsistencies to *encourage* diligence, while added negligible cognitive load to the annotation task.

Specifically, we included intentional errors in the dataset, duplicated some of the vibrations, and presented the existing annotations to the participants as "data from other users that can include noise". To identify the highly-biased participants or those who did not pay close attention to the experimental task, we included two intentional errors, one in the ratings and one in the tags, in each vibration group. For the rating

Fig. 5.5 Validation interface gave access to all 11 vibrations at the same time and could remove tags and adjust ratings. Participants could see the existing (expert) ratings in blue, and their own adjusted ratings in green. They could remove a tag by clicking on it (graying it out), and re-add it by clicking it again

error, we modified the energy$_d$ rating for one of the vibrations from very high ($+3$ on a 7-point likert scale) to very low (-3) or vice versa. For the tags, we added an invalid tag to one of the vibrations in each group (e.g., added "long" to a vibration with the short tag) resulting in two clearly contradicting tags for the vibration. These changes were clearly different from the characteristics and other ratings and tags for the vibration, thus added minimal cognitive load to the annotation task. Also, we duplicated one of the 10 vibrations in each group (for a total of 11 vibrations) to assess the participants' rating and tagging reliability.

Finally, as part of the experiment instructions, we told the participants that the existing ratings and tags were provided by other people and we were running this study to remove the noise in that data.

Apparatus and procedure: The validation interface was composed of two web pages, for calibration and annotation pages respectively (Fig. 5.5). An experiment session took about 1 h and the participants went over 2–3 vibration groups (22–33 vibrations) depending on their annotation speed. After the initial instructions, participants went through all the calibration vibrations for that session (33 vibrations). Then, they proceeded to the annotation page where they could see all the 11 vibrations for one group (randomized order). They could change the ratings, remove tags, or add additional tags; the initial ratings and tags were visible at all times. After completing a group, the experimenter loaded the next group of vibrations and the participant went through the calibration and annotation pages for that group. At the end of the session, participants filled a short post-questionnaire for demographic information and any other relevant comments.

5.4.3 Pre-processing of the Dataset

Prior to full analysis, we handled outliers and then averaged and incorporated our Stage 2 annotators' input to prune tags as planned.

Outlier removal: We used participants' performance on intentional rating and tag errors to identify outliers with high bias or low attention to the experimental task. Specifically, if a participant only modified the rating errors, we removed their tagging data and if they adjusted less than 1/3 of both the rating and tag errors, we removed all their data from the dataset. As a result, each vibration in the dataset has data from 9 taggers and 9–13 raters (5 rating outliers, and 13 tagging outliers).

Constructing the validated dataset: To derive the validated ratings for a vibration, we averaged all the participants' ratings for that vibration. We eliminated tags removed by more than 1/3 of the participants ($\geq$4 out of 9). In this way, we removed tags that were commonly marked as irrelevant, yet did not excessively limit the dataset (to the tags approved by everyone) to allow for more interesting analysis and results.

5.4.4 Definition of Analysis Metrics

To address our research questions, we devised a set of metrics that are applicable to ratings and free-form tags and used them as the basis for our analysis. Table 5.2 summarizes all the metrics with mathematical formulas. Below, V_i, $V_{i\prime}$ denote the ith vibration and its replica respectively. T_j refers to the jth tag, F_k to one of the four facets, and N_{items} to the number of items (e.g., tags, vibrations, participants). $\cap, \ominus$ denote the intersection and symmetric difference respectively of two tag sets.

5.5 Analysis and Results

We provide our analysis procedure and results, focusing on our three research questions in turn followed by a summary of our dataset characteristics (Table 5.4).

5.5.1 [Q1] Facet Substructure: What Are the Underlying Facet Dimensions That Dominate User Reactions to Vibrations?

To interpret and verify the underlying dimensions for the facets, we analyzed the data in four steps:

1. Ran a first MDS analysis on these vibration distances in each facet to determine the number of underlying dimensions for the facet;

Table 5.2 Definition of our analysis metrics

Tag removal threshold: The number of participants that must remove a tag from a vibration before we eliminate the tag in our validated dataset. For example, we use a tag removal threshold of 4, meaning that every tag that is removed by 4 or more participants from a vibration's list of tags is eliminated from the validated dataset

Vibration distance: The extent that two vibrations are described differently according to a given metric. In our study, the metrics are our facets. We calculate the distance between two vibrations in a facet (F_k) as the number of tags (N_t) that are different between the two vibrations divided by their total number of tags in the given facet. We use this metric in our MDS analysis of the vibrations.

$$Distance(V_i, V_j, F_k) = \frac{N_{tags}[(V_i, F_k) \ominus (V_j, F_k)]}{N_{tags}(V_i, F_k) + N_{tags}(V_j, F_k)} \tag{5.1}$$

Tag co-occurrence and tag distance: Co-occurrence is the number of times two tags are used together to describe the vibrations in our dataset. We calculate this value for two tags by counting the number of vibrations that have both tags and dividing it by their total frequency in our dataset.

$$Cooccurrence(T_i, T_j) = 1 - 2 \times \frac{N_{vibrations}(T_i \cap T_j)}{N_{vibrations}(T_i) + N_{vibrations}(T_j)} \tag{5.2}$$

Tag distance: We define distance between the two tags ("tag distance") as one minus their co-occurrence value. We use these tag distances in our MDS analysis on the tags.

$$Distance(T_i, T_j) = 1 - Cooccurrence(T_i, T_j) \tag{5.3}$$

Tag disagreement score: An estimate of the amount of controversy among the participants in keeping or removing a tag. We measure it based on the number participants that disagree with the majority of taggers (about removing or keeping a tag for a vibration) divided by the total number of times the tag is presented to the participants in our dataset. For example, if for all the occurrences of a tag in our dataset only one of the participant have a different opinion from the rest, the formula results in a disagreement score of 0.11. The highest disagreement is 0.44 (the lowest is 0) meaning that for all the vibrations, the tag is approved by half of the participants and removed by the other half.

$$Disagreement(T_i) = \sum_j \frac{N_{MinorityParticipants}(V_j, T_i)}{N_{vibrations}(T_i) \times N_{participants}(V_j)} \tag{5.4}$$

Vibration disagreement score: The amount of difference in the participants' descriptions of a vibration according to a criteria. In our study, we calculate vibration disagreement per rating and per facet. For the ratings, we use the standard deviation of the ratings by the participants. For each facet (i.e., tag set), we define our metric to be similar to the standard deviation but applicable to the tags. Specifically, for a vibration, we count the number of tags that are different between a participant's approved tags and the validated tag list for the vibration and divide it by total number of tags the experts provided for that vibration. We average the value over all taggers for that vibration.

$$Disagreement(V_i) = \sum_j \frac{N_{tags}[(V_i, P_j) \ominus (V_i, Validated)]}{N_{tags}(V_i, Experts)} \tag{5.5}$$

Unreliability score: *Rating* unreliability is the absolute difference in the ratings for a vibration and its duplicated version (for example, for energy ratings, the reliability is defined as $R(V_i, energy) = |energy(V_i) - energy(V_{l_i})|$). *Tag* unreliability is the percentage of removed tags that are different between a vibration and its replica. Specifically, it is the number of tags removed from a vibration or its replica (but not from both) divided by the total number of tags removed from each.

$$TagUnreliability(V_i, F_k) = \frac{N_{RemovedTags}[(V_i, F_k) \ominus (V_{l_i}, F_k)]}{N_{RemovedTags}(V_i, F_k) + N_{RemovedTags}(V_{l_i}, F_k)} \tag{5.6}$$

2. Determined an initial interpretation of the dimension semantics based on frequent and contrasting tags at the ends of each dimension (Table 5.3);
3. Visualized distribution of the vibrations along each MDS dimension, color-coded based on the existence (or lack) of related tags, to verify our interpretation of the dimension (Figs. 5.7, 5.8, 5.9, 5.10);
4. Examined results of a separate MDS analysis on tag (in contrast to vibration) distances as a test of convergent and discriminant validity (9.4).

Together, these analyses reinforced our interpretation of the semantics of the dimensions and revealed the distribution of vibrations and tags in each facet. Below, we separately describe the analysis steps in detail, then present results for each facet.

[Step 1] Deriving dimensions from vibration distances: We calculated quantitative values for vibration distances, in each facet, based on the the number of shared and different tags in the validated tag lists for each two vibrations in the library (Table 5.2). Then, we ran an MDS analysis on these vibration distance values for each facet. From this data, we determined the number of MDS dimensions using the eigenvalue plots as well as dimension interpretability. In Fig. 5.6, eigenvalue contributions are normalized to that of the first eigenvalue. Since these plots do not have an obvious

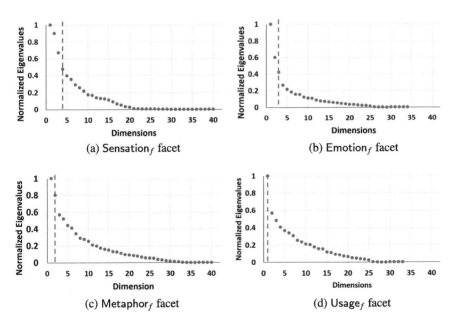

(a) Sensation$_f$ facet

(b) Emotion$_f$ facet

(c) Metaphor$_f$ facet

(d) Usage$_f$ facet

Fig. 5.6 Eigenvalue plots for the four facets. In each, the horizontal axis represents number of dimensions and the vertical axis indicates a dimension's contribution to reconstructing the vibration distances. If there is a large vertical gap between the nth and (n + 1)th dimensions, the first n dimensions have much larger contributions than the following ones and describe most of the variation in a facet. Thus, we use those first n dimensions in our analysis. The red dotted line highlights the number of dimensions we select for each facet. The eigenvalue contributions are normalized based on the first (largest) eigenvalue

Table 5.3 Final facet dimensions (derived in Table 5.4) and their most frequent tags: number of dimensions identified from MDS analysis on the vibration distances and our interpretation of their semantics (left column), most frequent tags and their rates of occurrence for the 10 vibrations at two ends of the dimensions (middle, right columns)

Dimension semantics	Negative end of scale	Positive end of scale
Sensation$_f$ Facet		
SensationD1:complexity$_d$	simple (8), regular (7), soft (7)	dynamic (10), irregular (9), complex (7)
SensationD2: continuity$_d$	discontinuous (10), regular (9)	continuous (10), simple (7)
SensationD3: roughness$_d$	smooth (10), soft (7), regular (7)	rough (8), short (6), discontinuous (6)
SensationD4: duration$_d$	discontinuous (7), simple (7), short (6)	grainy (8), regular (7), long (6), rough (6), ramping up (6)
Emotion$_f$ Facet		
EmotionD1: agitation$_d$	comfortable (10), calm (10), pleasant (8)	annoying (10), mechanical (9), agitating (9), urgent (9), angry (8)
EmotionD2: liveliness$_d$	predictable (10), boring (9), mechanical (9)	lively (10), unique (9), interesting (8), rhythmic (8)
EmotionD3: strangeness$_d$	rhythmic (10), lively (9), mechanical (8)	strange (10)
Metaphor$_f$ Facet		
MetaphorD1: on-off, nuanced/ongoing, repetitive$_d$	tapping (10)	engine (10)
MetaphorD2: natural/ mechanical$_d$	tapping (9), heartbeat (5)	alarm (10), game (7)
Usage$_f$ Facet		
UsageD1: urgency, attention-demand$_d$	pause (10), battery low (9), get ready (8), resume (7)	alarm (10), overtime (9), running out of time (9), above threshold (8)

"knee" (vertical gap), for each we first chose an initial set of dimensions based on the the highest-contributing eigenvalues; then, considered dimension interpretability before arriving at a final number [21]. We thereby found between one and four dimensions for each facet (Table 5.3).

[Step 2] Determine semantic descriptors for each MDS-produced dimension: We listed the validated tags and their rate of occurrence for the 10 farthest vibrations at each ends of an MDS dimension. The most frequent, yet still contrasting tags for the two ends of a dimension provided us with an initial interpretation of dimension semantics. We found one to several such *high-frequency tags* (descriptive terms) bounding each end of each dimension found in Step 1 (Table 5.3).

[Step 3] Verifying dimension semantics by visualizing vibration distributions: We visualized spatial distribution of vibrations along the identified MDS dimensions from Step 1 and color-coded them based on existence (red, green) or lack (gray)

Table 5.4 Facet dimension analysis

Sensation Facet

Dimensions from vibration distances: Figure 5.6a's eigenvalue plot suggests that after 4 primary dimensions, additional dimensions contribute little more (<0.1 apart). The identities of the most frequent tags at dimension extremes suggest that these four dimensions could be defined by their endpoints as: (1) simple/flat to complex/dynamic, (2) continuous to discontinuous, (3) smooth to rough, and (4) short to long (Table 5.3)

Color-coded vibration distributions: Figure 5.7 shows spatial distribution of the vibrations along the above four dimensions. All four have similar ranges (-0.5 to $+0.7$), indicating comparable variations along the dimensions. For the first three, the associated tags explain the dimension semantics well: green and red bars are well-separated at the two ends of the dimensions and the gray bars are around the central, neutral position. For the fourth dimension, the colored bars are less well separated, suggesting that these tags can at least partially explain this variation. We include it as the last interpretable dimension for the sensation$_f$ facet. These dimensions were further confirmed by our MDS analysis on tag distances (Sect. 9.4)

Final dimensions (also in Table 5.3): (1) simple—complex$_d$, (2) discontinuous—continuous$_d$, (3) smooth—rough$_d$, and (4) short—long$_d$. The overlap in the frequent tags for different dimensions (Table 5.3) and their spatial configuration (Fig. 9.1) suggest the above dimension properties are not completely orthogonal

Emotion Facet

Dimensions from vibration distances: Figure 5.6b's eigenvalues suggest 3–4 underlying dimensions; we opt for three due to higher interpretability. The most frequent tags in Table 5.3 suggest *(1) comfortable and calm versus annoying and urgent, (2) boring and predictable versus lively and interesting, (3) strange and surprising versus rhythmic and mechanical

Color-coded vibration distributions: Figure 5.8 shows the distribution of the vibrations along each emotion$_f$ dimension. For the first and second, color distribution follows our interpretation. For the last, green bars are mostly grouped at the right (strange and suprising) but red and gray bars are randomly dispersed on the left, suggesting the need for a better description for this end of the dimension.

Final dimensions: (1) comfortable—urgent, agitating$_d$, (2) boring—lively, interesting$_d$, (3) creepy, strange—rhythmic, predictable$_d$

Metaphor Facet

Dimensions from vibration distances: We removed 13 of 45 metaphor$_f$ tags that were applied with low frequency (to <2 vibrations) to avoid unrepresentative distortions in the MDS result. Metaphor$_f$'s eigenvalue plot then has a large number of dimensions with similar contributions; however, the first two slightly more so than others. Tag frequencies suggest that these two are differentiated in (1) tapping versus engine, and (2) tapping and heartbeat versus game or alarm. Further analysis of the tags, reported in 9.4, indicated that along dimension 1, tags are divided into ongoing and repetitive or pulse-like and nuanced. For dimension 2, tags tend to be natural and calm; or mechanical, synthetic and annoying (See Sect. 9.4 for the spatial configuration of tags)

Color-coded vibration distribution: Tag distributions for both dimensions show a separation of green and red bars at the two ends of the dimensions with gray bars lying mostly in the middle (Fig. 5.9).

Final dimensions: (1) on-off, nuanced—ongoing and repetitive$_d$ metaphors, and (2) natural, calm (mostly pulsing)—mechanical and annoying$_d$ metaphors.

Usage Facet

Dimensions from vibration distances: Eigenvalues suggest that the first dimension has a dominant contribution (Fig. 5.6d). According to the most frequent tags, this dimension represents urgency and attention-demand of notifications. On one end, usage$_f$ tags suggest time urgency while on the other, notifications require little attention and are mostly for users' awareness (Table 5.3)

Color-coded vibration distribution: In Figure 5.10, red, gray, and green bars are fairly well separated and gradually change from the left to the right of the dimension, supporting our one-dimension interpretation for the usage$_f$ facet

Final dimension: (1) Low-demand awareness—urgent and attention-demanding$_d$ notifications

---

of high-frequency tags from Step 2 (Figs. 5.7, 5.8, 5.9, 5.10). As explained more fully in the first caption, vertical bars encode MDS position of the vibrations along each dimension, while bar color denotes whether a vibration's validated tag list has one of that dimension's high-frequency tags. Red and green bars that are grouped at the opposite ends of the dimension with gray mostly in the middle signify that the identified tags adequately represent the semantics of the dimension; substantial mixing of colors does not.

[Step 4] Investigating tag distances: We ran a second MDS analysis on our derived *tag* distances (Table 5.2) and examined word positions in the resulting MDS map as a measure of convergent and discriminant validity of our interpretations [21], as follows. Convergent validity is supported when the words that have similar meanings in relation to a dimension are spatially close in the MDS solution. Discriminant validity is supported if the words with contrasting meanings are located far from each other in the MDS solution. Thus, we examined whether the contrasting tags for each dimension are far away from each other while the relevant tags for the dimensions are in the same area. Results from this step mainly support findings of the above steps and thus are reported in Sect. 9.4.

In Table 5.4, we step through this process to interpret the dimensionality of each of our facets specifically.

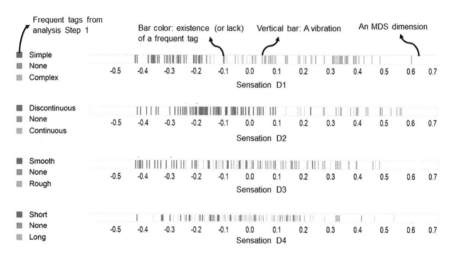

Fig. 5.7 Distribution of vibrations across the four MDS dimensions identified for the sensation$_f$ facet. All vibrations are shown. **Position coding**: Thin vertical bars project each vibration's MDS-derived location onto this dimension. **Color coding**: Bar color indicates whether the validated tag list for the vibration contains one of the frequent tags identified in Step 2 (red or green, with red indicating the left end of the dimension, and green the right end) or not (gray). For SensationD1$_d$, a red bar denotes that a vibration has a simple or a flat tag, while a green bar represents a vibration with a complex or dynamic tag and gray bars show vibrations with no related tag. SensationD2$_d$: (red:discontinuous, green:continuous), SensationD3$_d$: (red:smooth or soft, green:rough), SensationD4$_d$: (red:short, green:long)

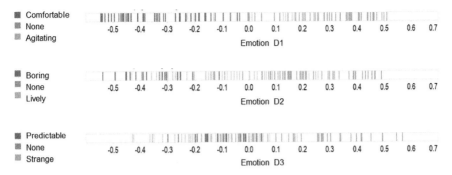

Fig. 5.8 Distribution of all the vibrations across the three MDS dimensions for the emotion$_f$ facet. EmotionD1$_d$: (red:calm, comfortable, or pleasant, green:urgent,annoying), EmotionD2$_d$: (red:boring, green:interesting, lively), EmotionD3$_d$: (red:predictable, familiar, green:strange, creepy, surprising)

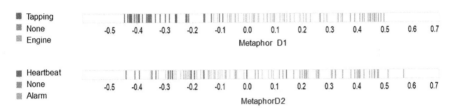

Fig. 5.9 Vibration distribution across the two MDS dimensions for the metaphor$_f$ facet. Tags for MetaphorD1$_d$: (red:tapping, green:engine), MetaphorD2$_d$: (red:heartbeat, green:alarm or game)

Fig. 5.10 Vibration distribution for the usage$_f$ facet. We color all vibrations with high urgency and attention tags (alarm, running out of time, overtime, or above intended threshold) with green marks; and red for those with awareness notifications (pause, battery low, resume, or get ready); and gray for those with none or a mix of both types

5.5.1.1 Our Five Rating Scales

To determine if our rating scales are orthogonal, we ran a Pearson correlation on the ratings for the five Likert-scale parameters across the 120 vibrations.

Results show significant medium to high correlation for all five parameters, except for pleasantness$_d$ and tempo$_d$ (low correlation, r $= -0.22$). Energy$_d$, arousal$_d$ and roughness$_d$ have the highest correlations (r $= 0.74 - 0.92$), followed by pleasantness$_d$ and roughness$_d$ (r $= -0.61$), tempo$_d$ with arousal$_d$ (r $= 0.56$), and roughness$_d$(r $= 0.52$), and pleasantness$_d$ with arousal$_d$ (r $= -0.53$) (full correlation table in Sect. 9.3).

5.5.2 [Q2] Between-Facet Linkages: How Are Attributes and Dimensions Linked Across Facets?

We address this question by examining linkages among our identified dimensions as well as linkages among the tags between various facets.

5.5.2.1 Dimension Level: Are There Linkages or Correlations Among the Identified Dimensions of Various Facets? What Factors can Describe These Correlations?

To address this question, we use factor analysis. Here, we include both the ratings and facet dimensions in our analysis to further link our derived facet structures to one another as well as to the rating metrics frequently used in the literature. Thus, our variables are the five rating scales and the 10 dimensions identified for all the facets (a total of 15 variables). We use the values of the 120 vibrations on those 15 variables as our samples. This results in a ratio of 8:1 for our analysis (8 samples per variable), satisfying the minimum suggested ratio in the literature (5:1) [48].

According to our results, four perceptual factors can describe correlations among the dimensions in various facets (the four right-most columns on Table 5.5). Table 5.5 shows the vibration properties (ratings and facet dimensions) with loadings >0.3 for each factor and highlights the high loadings (≥ 0.45) in **boldface**.

Factor 1 (Urgency$_{fact}$**):** UsageD1$_d$ and emotionD1$_d$ are highly connected to the same underlying factor as energy$_d$, arousal$_d$, roughness$_d$, and pleasantness$_d$. SensationD1 - complexity$_d$ and metaphorD2 - natural/mechanical$_d$ are also connected to this factor but with lower loadings.

Factor 2 (Liveliness/interestingness$_{fact}$**):** EmotionD2 - boring/lively$_d$ is connected with sensationD4 - duration$_d$, and tempo$_d$ on the second factor. SensationD2 - continuity$_d$ is also partially loaded onto this factor.

Factor 3 (Roughness$_{fact}$**):** This factor presents the link between sensationD3 - roughness$_d$ with roughness$_d$ and pleasantness$_d$ ratings.

Factor 4 (Novelty$_{fact}$**):** SensationD1 - complexity$_d$ and emotionD3 - strangeness$_d$ are connected on the fourth factor. MetaphorD1$_d$ also partially loads onto this factor.

5.5.2.2 Tag Level: How do Tags in the Different Facets Correlate?

We used our tag co-occurrence metric (Table 5.2) as a measure of correlation between tags in various facets. We report co-occurrence of the sensation$_f$ facet's tags with emotion$_f$, metaphor$_f$, and usage$_f$ tags, since sensation$_f$ tags more directly relate to engineering parameters (Fig. 5.11) but are also hardware independent. Figure 5.11 presents links among the emotion$_f$ and sensation$_f$ tags (see Sect. 9.6 for the tag co-occurrence tables of the metaphor$_f$ and usage$_f$ facets).

Table 5.5 Factor analysis outcome. The left column shows the initial rating scales† and new facet dimensions after MDS analysis. The next four columns present the factors upon which we have found some degree of cross-facet correlation, in terms of facet ratings and dimensions. For each factor column, **boldfaced** numbers highlight facet variables with the highest contributions to that factor (>0.45); empty cells indicate very low contributions (<0.3). Facet properties that have high values on the same factor column (e.g., energy, $UsageD1_d$ in the $Urgency_f$ factor) are correlated: the columns/factors are a point of linkages between the facets

Revised facet properties	Urgency (Factor 1)	Liveliness (Factor 2)	Roughness (Factor 3)	Novelty (Factor 4)
1. Sensation$_f$:				
energy$_d$†	**0.89**			
tempo/speed$_d$†	0.43	**0.45**	0.34	
roughness$_d$†	**0.75**		**0.48**	
SensationD1 - Complexity$_d$	**0.45**			**0.55**
SensationD2 - Continuity$_d$		−0.38		0.31
SensationD3 - Roughness$_d$			**0.89**	
SensationD4 - Duration$_d$	0.36	**−0.48**		
2. Emotion$_f$:				
pleasantness$_d$†	**−0.64**	0.33	−0.34	−0.31
arousal$_d$†	**0.95**			
EmotionD1 - Agitation$_d$	**0.82**			
EmotionD2 - Liveliness$_d$		**0.89**		
EmotionD3 - Strangeness$_d$				**0.60**
3. Metaphor$_f$:				
MetaphorD1 - On/off versus ongoing$_d$		−0.32		0.44
MetaphorD2 - Natural versus Mechanical$_d$	**0.45**			
4. Usage$_f$:				
UsageD1- Attention- demand$_d$	**0.80**			

Sensation tags

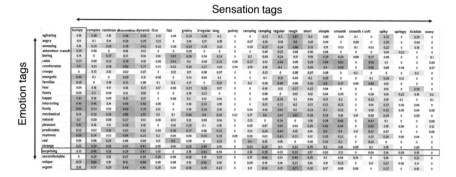

Fig. 5.11 Co-occurrence of the sensation$_f$ tags with the emotion$_f$ tags in our vibration library. For each emotion$_f$ tag (rows), we see the most (and least) associated sensation$_f$ tags (encoded as darker and lighter cells respectively). For example, highlighted with red boxes, to design a surprising vibration, one should make an irregular, dynamic, ramping up, and rough sensation (design scenario in Fig. 5.1a). Similarly, looking down on each column, one can see how a particular sensation$_f$ tag is perceived emotionally. Bumpy vibrations mostly invoke positive emotional response such as comfortable, energetic, happy, lively, etc. (evaluation scenario in Fig. 5.1b)

5.5.3 [Q3] Individual Differences: To What Extent Do People Coincide or Differ in Their Assessment of Vibration Attributes?

We examined variation in the participants' ratings and tags as a measure of individual differences in their perceptions and opinions. Here, we report these individual differences on various levels including the extent of variation (disagreement) across the facets, ratings, and tags as well as the amount of disagreement per vibration.

5.5.3.1 Per Facet

We measured overall individual differences in the facets based on percentage of facet tags that were approved by everyone (100% of the annotators), as well as percentage of tags that caused a split between the participants (defined as when half of participants removed a tag and the other half kept it as an appropriate tag for a vibration). Sensation$_f$ had the lowest individual differences, with the highest number of tags kept by everyone (21% compared to 7–12% for the other facets), and the lowest number of tags that caused a split (18% compared to 32–37%). Usage$_f$ elicited slightly more individuated responses than emotion$_f$ and metaphor$_f$, with 7% tags approved by everyone and 37% tags resulting in a split in the participants' opinions.

5.5.3.2 Per Rating

For each of the five rating scales, we used standard deviation of the values provided by all the annotators for a vibration as a measure of individual differences in that rating. Averaged across all vibrations and on a 7-point scale, these are 1.0, 0.8, 0.7, 0.7, 0.7 for pleasantness$_d$, roughness$_d$, energy$_d$, tempo$_d$, and arousal$_d$ respectively.

5.5.3.3 Per Tag

Stage 2 participants approved or removed some tags in consistent ways (e.g., short, irregular, agitating) whereas the participants showed differing opinions about the appropriateness of some others (e.g., ticklish, fear, start). *Tag disagreement score* represents the amount of controversy among the participants in keeping or removing a tag (Sect. 5.4.4). The highest possible score is 0.5, denoting a full split in participant opinions.

Figure 5.12 shows a bar chart of the number of tags in each facet, color-coded based on their disagreement score (higher color saturations denote higher disagreement scores). The figure also lists examples of tags with low and high disagreement scores: e.g., in sensation$_f$, short and smooth transition tags had the lowest disagreement while ticklish had the highest. Overall, usage$_f$ tags had higher disagreement compared to the other facets, with no tag showing very low (<0.2) disagreement.

5.5.3.4 Per Vibration

We computed disagreement among the ratings and tags assigned to each vibration (vibration disagreement score is defined in Sect. 5.4.4). Figure 5.13 presents a heatmap of a subset of vibrations and their disagreement scores for the ratings and tags (see disagreement values for all the vibrations in Fig. 9.6). Interestingly,

Fig. 5.12 A stacked bar chart showing tag disagreement scores in each facet. The height of each bar indicates total number of tags in a facet. More saturated parts of the bar indicate tags with higher disagreement scores

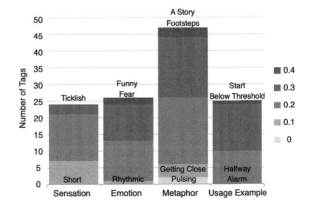

	Energy	Tempo	Roughness	Valence	Arousal	Sensation	Emotion	Metaphor	Usage
v-09-10-3-56	0.44	0.62	0.59	0.91	0.79	0.02	0.39	0.33	0.29
v-09-10-4-25	0.35	0.59	1.26	0.96	0.81	0.3	0.43	0.07	0.28
v-09-10-6-46	0.4	0.4	0.67	0.91	0.84	0.27	0.2	0.44	0.37
v-09-12-1-0	0.46	1.39	0.86	1.07	0.83	0.27	0.41	0.24	0.25
v-10-28-7-35	0.99	1.49	1.16	1.19	0.53	0.16	0.31	0.41	0.4
v-09-09-8-11	1.08	0.54	0.59	0.63	0.5	0.15	0.22	0.22	0.27
v-09-09-8-20	0.92	0.65	1.1	0.91	0.76	0.22	0.17	0.19	0.3
v-09-09-8-20-cpy	0.85	0.36	0.85	0.76	0.69	0.22	0.14	0.19	0.21
v-09-09-8-24	0.89	1.1	1.11	1.38	0.62	0.2	0.36	0	0.33
v-09-10-11-55	0.47	0.96	1.1	1.01	0.83	0.09	0.17	0.22	0.23
v-09-10-12-11	0.57	0.53	0.28	0.99	0.59	0.15	0.09	0.11	0.06
v-09-10-12-13	0.28	0.58	0.65	1.18	0.38	0.09	0.26	0.14	0.26
v-09-10-12-16	0.4	1.11	0.79	1.33	0.44	0.17	0.27	0.44	0.24
v-09-10-12-2	0.79	0.36	0.52	0.86	0.28	0.16	0.2	0.22	0.22
v-09-10-12-6	0.77	0.62	0.46	0.93	1.01	0.03	0.17	0.16	0.27
v-09-10-12-9	0.53	0.5	0.66	0.88	0.18	0.21	0.26	0.27	0.28
v-09-10-12-9-cpy	0.68	0.33	0.73	1.11	0.66	0.24	0.19	0.38	0.39
v-09-10-3-52	0.14	0.26	0.57	0.83	0.64	0.2	0.28	0.24	0.22

Fig. 5.13 Disagreement scores for the ratings and facets for a subset of the vibrations, calculated based on Table 5.2. Disagreement scores are within [1–7] (ratings), and [0,1] (facets). A vibration can have a low disagreement score on one rating or tag set but a high disagreement score on another. High saturation denotes high disagreement

the vibrations were not always consistently disagreed or agreed upon. For example, vibration "v-09-10-3-56" had low disagreement on sensation$_f$ tags but higher disagreement on emotion$_f$, metaphor$_f$, and usage$_f$ tags. The vibrations also differed in the facet(s) that had the *lowest* controversy for them: "v-09-10-6-46" was mostly agreed upon in the emotion$_f$ facet but had high disagreement in the metaphor$_f$ facet. This pattern was reversed for another vibration (e.g., "v-09-10-4-25").

5.5.4 Methodology: How Does Staged Data Collection Impact Annotation Quality?

The goal of our two-stage data collection was to reduce noise from outliers and improve dataset convergence and reliability by facilitating the annotation task for the lay users, but at the cost of an additional round of data collection. Below, we summarize how well this new method achieves these goals by examining dataset characteristics after the two rounds of annotations and reliability of the final dataset. *Expert and Lay User Annotations*: Table 5.6 summarizes characteristics of our dataset after expert and lay user annotation stages.
Reliability of the final annotation set: To assess reliability, we measured absolute rating difference and percentage of tag difference between a vibration and its replica (Sect. 5.4.4) for each individual participant as well as for the final aggregated dataset. On average, the ratings were ∼0.7 apart (on a 7-point scale) for individual participants but this difference was reduced to ∼0.2 for the final aggregated dataset. Further,

Table 5.6 Summary of our annotation dataset after the two stages of expert annotation and lay user validation (i.e., pruning). The left column indicates: the the average difference in values provided on the five rating scales originally used to define the facets (top section); overlap in the tag sets for each of the facets (middle section); and the overall tag count for these facets (bottom section). Values in the experts and lay user columns in Table 5.6 cannot be directly compared due to differences in the tasks in these collection stages: experts applied annotations (each vibration was annotated by two of three experts), while lay users were asked to confirm them, and largely removed rather than adding tags

	Experts	**Lay Users**
	Average difference among experts	**Average deviation from experts**
Rating difference	**(Range, 7-point scale)**	**(Range, 7-point scale)**
Energy$_d$	1.15	0.45
Tempo$_d$	1.26	0.54
Roughness$_d$	1.6	0.64
Pleasantness$_d$	1.64	0.84
Arousal$_d$	1.5	0.5
Tag overlap	**Tags applied by both experts**	**Tags approved by $\geq$ 4 lay users**
Sensation$_f$	25%	86%
Emotion$_f$	17%	72%
Metaphor$_f$	14.5%	76%
Usage$_f$	12.5%	69%
Dataset tag count	**Following expert annotation**	**Following lay-user validation**
Sensation$_f$	744	635
Emotion$_f$	988	716
Metaphor$_f$	584	442
Usage$_f$	1234	857

$\sim$33% of the tags removed by an individual were different between a vibration and its replica which was further reduced to $\sim$7% difference on the final aggregated set.

5.6 Discussion

We start by looking at how these results apply to the three design, evaluation, and personalization scenarios we proposed in the introduction (Fig. 5.1): have we indeed found evidence for *perceptually continuous dimensions within individual facets, along which users would presumably find it logical to "move" individual haptic elements as an act of design?* Do we have a mapping among the facets that enables translation of design requirements, or evaluation of aesthetic properties of haptic elements?

We then compare our facet dimensions with the perceptual vibrotactile properties in the literature and draw insights into findings on individual differences and annotation reliability. We finish by reviewing the validity and effectiveness of our methodological choices.

5.6.1 Within-Facet Perceptual Continuity: Scenarios

Scenario 1—Design Guidelines and Manipulations (Fig. 5.1a): In making haptic sensations, designers commonly have a set of requirements in the usage$_f$, metaphor$_f$, or emotion$_f$ facets (e.g., surprise or racing car engine) and require guidelines prescribing important sensation$_f$ or engineering parameters for meeting those requirements. The linkages between the facets can provide such guidelines: the designer can look along the rows of Fig. 5.11 and find the highly correlated sensation$_f$ tags. For example, using Fig. 5.11, the task of designing a surprise vibration is broken into designing a sensation that is irregular, complex, ramping up, and rough (sensation$_f$ tags with high co-occurrence with surprise).

On the dimensional level, between-facet linkages provide a more continuous mapping for design. For example, a designer might want to create a palette of sensations that vary in liveliness. Using the correlation among the boring—lively$_d$ dimension and the dimensions from the sensation$_f$ facet, the designer can vary continuity$_d$ and tempo$_d$ of the vibrotactile rhythm in sketching alternative palettes for further investigation.

Determining the relevant engineering parameters and their values depends on the actuator type (e.g., voice coil vs. eccentric rotating mass actuators), and its hardware configurations (e.g., form factor, weight) and is straightforward, given the body of psycho-physical and sensory studies in haptics. For example, the designer can add discontinuity by including silence or pause in a vibration while ensuring that the duration of silence is perceptible to people [18].

Scenario 2—Evaluation (Fig. 5.1b): Alternatively, for cases where a designer has a set of vibrations and is interested to know their emotional connotations, proper metaphors or usage examples, he/she can look them up along the columns of Fig. 5.11. For example, a bumpy sensation usually has positive emotional connotations such as happy, interesting, lively and rhythmic, while ramping up sensations are usually annoying, mechanical, and uncomfortable.

Scenario 3—Personalization (Fig. 5.1c): Facet dimensions and their linkages provide the theoretical grounding for designers to *build tuning and stylization tools* for end-users who may wish to personalize their vibration notifications. First, the dimensions we found in this chapter are good candidates for being the basis of tuning sliders, as they capture the dominant spectrums along which a vibration can vary in a facet. For example, one can imagine a tuning slider that moves a vibration along the emotion dimension of boring—lively$_d$. Then, even more practically, the linkages, identified in our results, between a dimension in the emotion$_f$, metaphor$_f$, and usage$_f$ facets and the sensation$_f$ dimensions inform us about the mechanics of building these sliders. For example, the boring—lively$_d$ dimension is correlated with the signal's tempo, duration$_d$ (sensationD4) and continuity$_d$ (sensationD2). Thus, a designer can use these three sensation$_f$ attributes in developing an algorithm for a liveliness slider, which is ultimately controlled by end-users to modify a vibration's liveliness for their personal taste. In Chap. 7, we use these results to build a set of tuning sliders for vibrations.

5.6.2 Facet Dimensions and Linkages

Here, we discuss the unique insights and challenges for the facet dimensions and present implications for future research and design when applicable.

Sensation$_f$ provides designers with a practical translation platform between the facet space and engineering parameters like frequency and waveform Sensation$_f$ dimensions reflect important perceptual and engineering parameters identified in past studies. Specifically, *rhythm* and *envelope*, two parameters found to be influential and manipulable in expressive vibrotactile design [18, 51], are directly linked to continuity$_d$ and complexity$_d$ (sensationD2, D1 respectively). Roughness$_d$ and duration$_d$ are also known to impact users' perception [42, 51, 52]. Thus, translating the emotion$_f$, metaphor$_f$, and usage$_f$ dimensions and tags to the sensation$_f$ facet offers a practical and hardware-independent means for design.

Emotional perceptions of vibrations do not follow theoretical dimensions of pleasantness$_d$ and arousal$_d$. Correlation of the pleasantness$_d$ and arousal$_d$ ratings (Sect. 5.5.1.1) as well as our MDS results on the emotion$_f$ tags suggest that these two dimensions are not orthogonal for our vibrotactile collection. As a result, not all four quadrants of the pleasantness (valence)-arousal grid are covered by the vibrotactile sensations in our library. Specifically, none were marked as either very pleasant and alarming (positive valence-positive arousal), or very calm but unpleasant (negative valence-negative arousal).

While it is possible that such examples exist but our library does not contain them, we note that two recent studies found a similar correlation and also the same gap for different vibrotactile actuators and vibration sets. Yoo et al. examined several sets of vibrations (24–36 items each) on a voice coil actuator (Haptuator—[53]) and none covered the negative valence-negative arousal or very high valence-high arousal quadrants [54]. Our own previous study in Chap. 2 reports a similar correlation for a small subset of 14 vibrations on an Electro-Active Polymer (EAP) actuator.

We propose that for vibrations, the theoretical dimensions of pleasantness$_d$ and arousal$_d$ in the literature are not good representatives for the 2-D affect grid. There, sad and boring have negative valence and negative arousal while vibrations with sad and boring tags do not fall in that area; they are not necessarily unpleasant and quiet and this difference is reflected in our dataset. Instead, our MDS analysis on the emotion$_f$ tags suggest that people perceive and rate vibrations according to three other dimensions: (1) agitation$_d$, (2) liveliness$_d$, and (3) strangeness$_d$.

This result impacts future research and design in at least three ways. First, further studies are needed to *confirm or reject this pattern* using other vibration sets, and compare emotion$_f$ dimensions for vibrations with other haptic stimuli (such as natural textures, force feedback and variable friction) and other modalities such as vision and audition. Each of these stimuli categories have distinct similarities and differences with vibrotactile sensations, impacting users' emotional experience (e.g., variable friction stimuli are primarily sensed through skin but require active user movement). Thus, future research is required to examine their emotional space(s) and contrast them with our proposed emotion space for vibrotactile sensations. Second, the three dimensions provide *new directions for vibration design*. Agitation$_d$, liveliness$_d$, and strangeness$_d$ explain large variations in emotion$_f$, have low correlation, and provide a more accessible design space for current vibrotactile technology. They may be promising targets for affective design. Finally, once further validated, these dimensions offer good candidates for devising a standard *evaluation instrument* for vibrations.

Metaphor$_f$ dimensions are the most difficult to interpret. Our results suggest two dimensions for metaphor$_f$ tags that vary on continuity, novelty, and urgency. However, the spatial configuration of tags in Fig. 9.3 does not completely follow this definition (see the report of outlier metaphor$_f$ tags in Fig. 9.3). Also, these two dimensions are partially linked to the other facets in our factor analysis. One reason could be that our metaphor$_f$ tag set is larger but also sparser: there are fewer common metaphor$_f$ tags among the vibrations (Table 5.3) compared to sensation$_f$, emotion$_f$, and usage$_f$ tags. While this trend can reflect an inherent characteristic of metaphors for describing vibrations, future studies are needed to validate and expand on the above dimensions and further develop the metaphor$_f$ vocabulary for vibrotactile effects.

Users' interpretation of vibration meaning in usage contexts is mainly dictated by their energy (or urgency). According to our MDS results, vibration energy$_d$ or urgency$_d$ is the most important dimension for usage$_f$ tags. While energy is an important design parameter, we are not aware of previous work that empirically connects a vibration's energy to its application. Our vibration library is designed to include a wide range of sensations but our tag list for usage$_f$ is developed for a specific context: applications where time tracking is an important component (e.g., giving presentations and exercising). We anticipate this finding to extend to other application contexts but future studies are needed to confirm or reject the importance of energy for other types of applications.

Emotional connotations of vibrations play an important role in users' perception of vibrations, regardless of facet. The three dimensions found for emotion$_f$ have substantially high loadings on three of the four factors in Table 5.5: urgency$_{fact}$, liveliness$_{fact}$ and novelty$_{fact}$. This suggests that the underlying constructs, describing the variations and linkages between the facets, are mainly emotional. In the absence of other strong criteria, the emotion$_f$ facet can serve as the best default for end-user tools and interfaces.

5.6.3 Individuals' Annotation Reliability and Variation

Reliability of individuals' tagging is surprisingly low. In our Stage 2 study component, we placed a duplicate vibration in each vibration set—i.e., two out of the 11 were identical (Sect. 5.4.2). However, about 33% of individuals' removed tags differed for these duplicates (Sect. 5.5.4). This number is unexpectedly high: participants had access to all the vibrations and their tags via the experiment interface. Although the variation may be partially due to varying commitment and focus, it also suggests that people's memory of vibrations quickly fades. In contrast to auditory and visual icons, sensations in this unfamiliar modality are not always immediately memorable, and users commonly play a vibration several times to form an opinion about it or to compare it with another vibrotactile sensation. This negatively impacts reliability, but in some cases can simplify study design when one stimulus is presented in multiple experimental conditions.

Data on individual differences in ratings and tags inform haptic evaluation. Disagreement scores for the tags and ratings suggest that a notable portion of annotation variation is due to differences among users' definitions of the language terms and its manifestation in a tactile signal. This is evidenced by lower individual difference values for sensation$_f$ tags and the five rating scales. To mitigate this in the long run, we need to devise and consistently use a set of standard rating scales; the facet dimensions are promising candidates for such an endeavor. In the meantime, our tag disagreement scores can inform haptic researchers in selecting less controversial tags or estimating the number of participants required for their evaluation.

5.6.4 Review of Our Methodology

We contribute a data collection and analysis methodology, based on existing practices in the music annotation domain, that allows for comprehensive evaluation of a large vibration collection. Here, we discuss the validity and effectiveness of our methodological choices according to our results to support future uses and adaptations of our approach.

5.6.4.1 Method Validity

Bias in validation stage: Seeing existing annotations did not override participant perceptions. Participants made large adjustments ($\sim$4.3 on a 7-point scale) to the intentional energy rating errors applied in the validation stage to identify outliers—Sect. 5.4.2). Also, a notable percentage of the tags ($\sim$14–31%) are removed by 4 or more (out of 9) participants, demonstrating some degree of inter-participant consistency as well as willingness to respond with initiative. We also guarded against bias by describing the existing annotations to the participants as "noisy data from other users;" and by eliminating the participants with few annotation adjustments as outliers, on presumption that this indicated low engagement with the task. Finally, our validation task resembles practical scenarios where users start from a proposed set of notifications and their intended perception and usage (e.g., list of alarm tones on a phone, game sounds, etc.) and adopt or reject notifications depending on their perceptual match. Thus, although we expect some degree of conformity among the participants to the existing tags and ratings which were their (nonzero) starting point, it appears this did not override their choices and our validated dataset reflects their accepted annotations among the proposed ones.

Annotation instrument: Quality of our tag lists are reflected in the resulting facet dimensions. While developing the tag sets, our goal was to include as many relevant tags as possible, yet avoid redundant tags. For sensation$_f$ and emotion$_f$, our tag lists were built on existing adjective lists in the literature, were inclusive and were independent of the context. Thus, for these facets we could identify several dimensions with stronger linkages in the factor analysis. In contrast, the metaphor$_f$ and usage$_f$ tag lists were use-case dependent and could not be inclusive in nature. Further, it was more difficult to identify tag redundancy and conflicts for them. Thus, they resulted in fewer dominant dimensions which were harder to interpret (metaphor$_f$) and more use case dependent (usage$_f$). The attributes and dimensions for these facets can be further refined and validated over time, through follow up studies that examine other use cases and metaphors with diverse participants.

Future work can further refine our metaphor$_f$ and usage$_f$ attributes and dimensions by studying other use cases and participant groups.

Analysis methods: We triangulate our analysis to guard against the subjectivity in our interpretations. For both MDS and factor analysis, researchers determine number and semantics of dimensions and factors. Although this interpretation is based on evidence in the data, the resulting semantics are subject to the researchers' bias and pre-conceptions. To guard against this, we use three different analyses on the tags to interpret semantics of the facet dimensions and provide data on between-facet linkages on both dimensional and tag level.

Analysis methods: Factors with low loadings must be interpreted with caution. Our factor analysis has a ratio of 8:1 for data points (120 vibration ratings and MDS positions) and variables (15 ratings and facet dimensions; Sect. 5.5.2.1). While this meets the minimum ratio proposed in the literature (5:1), higher ratios (10:1 or more) are recommended for more stable results [48]. With our data, the variables with low factor loadings may not be stable if more data is added, thus they must be regarded with caution. This is especially true for the two metaphor$_f$ dimensions and for continuity$_d$ (sensationD2).

5.6.4.2 Method Effectiveness

Recruitment benefits: The staged approach increases efficiency of data collection and improves convergence. Practically speaking, we found that validating existing ratings and tags can be done more quickly than annotating a vibration. In our study, validation sessions include about three times more vibrations than our pilot and expert annotation sessions (33 vibrations compared to 12 vibrations). This means the same amount of data can be collected with fewer participants. Further, we found that the between-subject variations in the validation stage were reduced to values equal to within-subject variations (reliability) in the ratings, leading to better convergence. In Sects. 5.5.3.2 and 5.5.4, all values are ≤1 on a 7-point Likert scale. Finally, having expert ratings on the vibrations allowed for quick detection of outliers in the data and adjusting the recruitment plan accordingly.

Value added by end-user validation: Second stage is crucial for validating expert tags. On average, the lay-user-validated ratings are about 0.5 (7-point scale) different from the expert ratings, and the lay-user-validated set of tags include 14–31% fewer tags than the expert tag set. These results suggest that in this study experts' ratings provide a fairly accurate estimate of users' ratings; while for the tags, experts' and lay participants' opinions deviate more, justifying the need for the validation stage. If further studies confirm this pattern, then this approach can provide a *discount evaluation method* for vibrotactile design similar to heuristic evaluation in user interface design [55].

5.7 Conclusion

In this chapter, we presented four vibration facets, their underlying dimensions and their linkages and mappings based on ratings and tags collected for a library of 120 vibrations; Fig. 5.3 illustrates the emergent landscape we have exposed and described with tags, facets, dimensions and facet-linking factors. Our data and analysis confirmed definite cross-facet linkages between certain facet dimensions. We described these linkages on a discrete level between tags (descriptive words applied to specific vibrations, which themselves we empirically located within facet dimensional space) and on a continuous level between dimensions$_d$ (wherein dimensions provide perceptual delineation of the facets). For the latter, the linkages were described according to four factors (perceptual constructs underlying facet linkages): a vibration's urgency$_{fact}$, liveliness$_{fact}$, roughness$_{fact}$ and novelty$_{fact}$.

The linkages between the sensation$_f$ facet and the other facets (on both tag and dimension levels) offer guidelines for vibration design, evaluation, and personalization. However, we still lack a continuous mapping between most facet parameters (user's cognitive schemas) and the engineering parameters, by which these sensations are constructed. Applying machine learning techniques to the vibratory signals and their associated disposition within the facet space (such as the ratings, tags and MDS positions on the facet dimensions) is one approach towards identifying such a mapping. To this end, we have released our vibration dataset (vibration .wav files, their annotations and MDS characterization) for use by other researchers [26].

In the next chapter, we examine this mapping in the use case of developing a set of tuning sliders that can move a vibration along the semantic facet dimensions—that is, Scenario 3.

Will underlying facet dimensions and linkages apply to sensations produced with other haptic technologies? We anticipate that to a large extent they will, although specific labels and properties for the facets might vary. The literature includes evidence that people use sensation$_f$, emotion$_f$, and metaphor$_f$ descriptions for many kinds of haptic sensations, ranging from ultrahaptics effects (non-contact stimuli produced with acoustic waves [11]) to movements of a furry touch-based social robot [56, 57]. Confirming this requires future studies that examine the facet dimensions for other types of haptic sensations, such as force feedback, texture displays, variable friction and ultrahaptics, and comparing their findings with our results. Such an endeavor can lead to a more holistic and technology-independent model of user haptic perception.

We close this chapter by noting that rarely have the many challenges inherent in haptic evaluation [1] been approached through the development of new, haptic-specific methodologies and evaluation instruments. In this chapter, we offered a novel, scalable data collection approach to mapping users' comprehension of large sets of haptic signals; and report between- and within-subject data variation that can inform future instrument development.

References

1. MacLean, K.E., Schneider, O., Seifi, H.: Multisensory haptic interactions: understanding the sense and designing for it. In: The Handbook of Multimodal-Multisensor Interfaces. ACM Books (2017)
2. Lo, J., Johansson, R.S., et al.: Regional differences and interindividual variability in sensitivity to vibration in the glabrous skin of the human hand. Brain Res. **301**(1), 65–72 (1984)
3. Hollins, M., Bensmaïa, S., Karlof, K., Young, F.: Individual differences in perceptual space for tactile textures: evidence from multidimensional scaling. Percept. Psychophys. **62**(8), 1534–1544 (2000)
4. Peck, J., Childers, T.L.: Individual differences in haptic information processing: the need for touch scale. J. Consum. Res. **30**(3), 430–442 (2003)
5. Levesque, V., Oram, L., MacLean, K.E.: Exploring the design space of programmable friction for scrolling interactions. In: Proceedings of IEEE Haptic Symposium (HAPTICS '12), pp. 23–30 (2012)
6. Chan, A., MacLean, K., McGrenere, J.: Designing haptic icons to support collaborative turn-taking. Int. J. Hum.-Comput. Stud. (IJHCS) **66**(5), 333–355 (2008)
7. Tam, D., MacLean, K.E., McGrenere, J., Kuchenbecker, K.J.: The design and field observation of a haptic notification system for timing awareness during oral presentations. In: Proceedings of the ACM SIGCHI Conference on Human Factors in Computing Systems (CHI '13), pp. 1689–1698. ACM, New York (2013). https://doi.org/10.1145/2470654.2466223
8. Zhao, S., Schneider, O., Klatzky, R., Lehman, J., Israr, A.: Feelcraft: crafting tactile experiences for media using a feel effect library. In: Proceedings of the Adjunct Publication of the 27th Annual ACM Symposium on User Interface Software and Technology (UIST '14), pp. 51–52. ACM, New York (2014). https://doi.org/10.1145/2658779.2659109
9. Swindells, C., Pietarinen, S., Viitanen, A.: Medium fidelity rapid prototyping of vibrotactile haptic, audio and video effects. In: Proceedings of IEEE Haptics Symposium (HAPTICS '14), pp. 515–521 (2014)
10. Israr, A., Zhao, S., Schwalje, K., Klatzky, R., Lehman, J.: Feel effects: Enriching storytelling with haptic feedback. ACM Trans. Appl. Percept. (TAP) **11**, 11:1–11:17 (2014)

11. Obrist, M., Seah, S.A., Subramanian, S.: Talking about tactile experiences. In: Proceedings of the ACM SIGCHI Conference on Human Factors in Computing Systems (CHI '13), pp. 1659–1668. ACM (2013)

12. Schneider, O.S., MacLean, K.E.: Improvising design with a haptic instrument. In: Proceedings of IEEE Haptics Symposium (HAPTICS '14), pp. 327–332. IEEE (2014)

13. Yee, K.P., Swearingen, K., Li, K., Hearst, M.: Faceted metadata for image search and browsing. In: Proceedings of the ACM SIGCHI conference on Human Factors in Computing Systems (CHI '03), pp. 401–408 (2003)

14. Smith, G., Czerwinski, M., Meyers, B., Robbins, D., Robertson, G., Tan, D.S.: Facetmap: a scalable search and browse visualization. IEEE Trans. Vis. Comput. Graph. **12**(5), 797–804 (2006)

15. Hearst, M.: Design recommendations for hierarchical faceted search interfaces. In: Proceedings of the ACM SIGIR Workshop on Faceted Search, pp. 1–5 (2006)

16. Hearst, M.A.: UIs for faceted navigation: Recent advances and remaining open problems. In: Proceedings of the Second Workshop on Human-Computer Interaction and Information Retrieval (HCIR), pp. 13–17 (2008)

17. Fagan, J.C.: Usability studies of faceted browsing: a literature review. Inf. Technol. Libr. **29**(2), 58 (2010)

18. Ternes, D.R.: Building large sets of haptic icons: Rhythm as a design parameter, and between-subjects mds for evaluation. Ph.D. thesis, The University of British Columbia (2007)

19. Seifi, H., Zhang, K., MacLean, K.E.: Vibviz: organizing, visualizing and navigating vibration libraries. In: Proceedings of IEEE World Haptics Conference (WHC '15), pp. 254–259. IEEE (2015)

20. Ternes, D., Maclean, K.E.: Designing large sets of haptic icons with rhythm. In: Haptics: Perception, Devices and Scenarios, pp. 199–208. Springer, Berlin (2008)

21. Guest, S., Dessirier, J.M., Mehrabyan, A., McGlone, F., Essick, G., Gescheider, G., Fontana, A., Xiong, R., Ackerley, R., Blot, K.: The development and validation of sensory and emotional scales of touch perception. Atten. Percept. Psychophys. **73**(2), 531–550 (2011)

22. The NounProject, Inc.: (2016). http://thenounproject.com/. Accessed 24 July 2016

23. Turnbull, D., Barrington, L., Lanckriet, G.R.: Five approaches to collecting tags for music. Proc. Int. Soc. Music Inf. Retr. (ISMIR) **8**, 225–230 (2008)

24. Cox, T.F., Cox, M.A.: Multidimensional Scaling. CRC Press, Boca Raton (2000)

25. Thompson, B.: Exploratory and confirmatory factor analysis: Understanding concepts and applications. American Psychological Association (2004)

26. Seifi, H., MacLean, K.E.: VibViz Dataset (2016). http://www.cs.ubc.ca/labs/spin/vibviz. Accessed 29 July 2016

27. Brunet, L., Megard, C., Paneels, S., Changeon, G., Lozada, J., Daniel, M.P., Darses, F.: Invitation to the voyage: The design of tactile metaphors to fulfill occasional travelers' needs in transportation networks. In: IEEE World Haptics Conference (WHC '13), pp. 259–264 (2013). https://doi.org/10.1109/WHC.2013.6548418

28. Zheng, Y., Su, E., Morrell, J.B.: Design and evaluation of pactors for managing attention capture. In: Proceedings of IEEE World Haptics Conference (WHC '13), pp. 497–502 (2013)

29. Ryu, J., Choi, S.: posVibEditor: Graphical authoring tool of vibrotactile patterns. In: Proceedings of IEEE International Workshop on Haptic Audio visual Environments and Games (HAVE), pp. 120–125 (2008)

30. Schneider, O.S., MacLean, K.E.: Studying design process and example use with macaron, a web-based vibrotactile effect editor. In: Proceedings of IEEE Haptics Symposium (HAPTICS '16), pp. 52–58 (2016)

31. Hong, K., Lee, J., Choi, S.: Demonstration-based vibrotactile pattern authoring. In: Proceedings of the Seventh International Conference on Tangible, Embedded and Embodied Interaction (TEI '13), pp. 219–222 (2013)

32. Schneider, O.S., Israr, A., MacLean, K.E.: Tactile animation by direct manipulation of grid displays. In: Proceedings of the 28th Annual ACM Symposium on User Interface Software and Technology (UIST '15), pp. 21–30. ACM (2015)

33. Lieberman, H., Paternò, F., Klann, M., Wulf, V.: End-user development: An emerging paradigm. In: End User Development, pp. 1–8. Springer, Berlin (2006)
34. Saul, G., Lau, M., Mitani, J., Igarashi, T.: Sketchchair: an all-in-one chair design system for end users. In: Proceedings of the Fifth ACM International Conference on Tangible, Embedded, and Embodied Interaction (TEI '11), pp. 73–80 (2011)
35. Evening, M.: The Adobe Photoshop Lightroom 5 Book: The Complete Guide for Photographers. Pearson Education, London (2013)
36. Harrower, M., Brewer, C.A.: Colorbrewer.org: an online tool for selecting colour schemes for maps. Cartogr. J. (2013)
37. van Erp, J.B., Spapé, M.M.: Distilling the underlying dimensions of tactile melodies. Proc. Eurohaptics Conf. **2003**, 111–120 (2003)
38. MacLean, K., Enriquez, M.: Perceptual design of haptic icons. In: Proceedings of EuroHaptics Conference, pp. 351–363 (2003)
39. Okamoto, S., Nagano, H., Yamada, Y.: Psychophysical dimensions of tactile perception of textures. IEEE Trans. Haptics (ToH) **6**(1), 81–93 (2013)
40. Doizaki, R., Watanabe, J., Sakamoto, M.: A system for evaluating tactile feelings expressed by sound symbolic words. In: Auvray, M., Duriez, C. (eds.) Haptics: Neuroscience, Devices, Modeling, and Applications: Proceedings of Eurohaptics, pp. 32–39. Springer, Berlin (2014). https://doi.org/10.1007/978-3-662-44193-0_5
41. Brown, L.M., Brewster, S.A., Purchase, H.C.: Tactile crescendos and sforzandos: applying musical techniques to tactile icon design. In: CHI'06 Extended Abstracts on Human factors in Computing Systems (CHI EA '06), pp. 610–615. ACM (2006)
42. Hoggan, E., Brewster, S.: Designing audio and tactile crossmodal icons for mobile devices. In: Proceedings of the 9th ACM International Conference on Multimodal Interfaces (ICMI '07), pp. 162–169. ACM (2007)
43. Grey, J.M.: Multidimensional perceptual scaling of musical timbres. J. Acous. Soc. Am. **61**(5), 1270–1277 (1977)
44. Pandora Internet Radio: Pandora (2016). http://www.pandora.com/. Accessed 24 July 2016
45. Jäschke, R., Marinho, L., Hotho, A., Schmidt-Thieme, L., Stumme, G.: Tag recommendations in folksonomies. In: Proceedings of European Conference on Principles of Data Mining and Knowledge Discovery, pp. 506–514. Springer, Berlin (2007)
46. Last.fm: (2016). http://www.last.fm/music. Accessed 24 July 2016
47. The MathWorks, Inc.: Matlab (2016). https://www.mathworks.com/products/matlab.html. Accessed 29 July 2016
48. Yong, A.G., Pearce, S.: A beginner's guide to factor analysis: focusing on exploratory factor analysis. Tutor. Quant. Methods Psychol. **9**(2), 79–94 (2013)
49. Jason L. Huang Paul G. Curran, J.K.E.M.P.R.P.D.: Detecting and deterring insufficient effort responding to surveys. J. Bus. Psychol. **27**(1), 99–114 (2012). http://www.jstor.org/stable/41474909
50. Curran, P.G.: Methods for the detection of carelessly invalid responses in survey data. J. Exp. Soc. Psychol. **66**, 4–19 (2016). https://doi.org/10.1016/j.jesp.2015.07.006. http://www.sciencedirect.com/science/article/pii/S0022103115000931. Rigorous and Replicable Methods in Social Psychology
51. MacLean, K.E.: Foundations of transparency in tactile information design. IEEE Trans. Haptics (ToH) **1**(2), 84–95 (2008)
52. Hoggan, E., Brewster, S.: New parameters for tacton design. In: CHI'07 Extended Abstracts on Human Factors in Computing Systems (CHI EA '07), pp. 2417–2422. ACM, New York (2007)
53. TactileLabs: (2016). http://tactilelabs.com/. Accessed 29 July 2016
54. Yoo, Y., Yoo, T., Kong, J., Choi, S.: Emotional responses of tactile icons: Effects of amplitude, frequency, duration, and envelope. In: Proceedings of IEEE World Haptics Conference (WHC'15), pp. 235–240 (2015). https://doi.org/10.1109/WHC.2015.7177719
55. Nielsen, J., Molich, R.: Heuristic evaluation of user interfaces. In: Proceedings of the ACM SIGCHI Conference on Human Factors in Computing Systems (CHI '90), pp. 249–256 (1990)

56. Yohanan, S., MacLean, K.E.: Design and assessment of the haptic creature's affect display. In: Proceedings of the Sixth ACM International Conference on Human-Robot Interaction (HRI '11), pp. 473–480 (2011)
57. Yohanan, S., Chan, M., Hopkins, J., Sun, H., MacLean, K.: Hapticat: exploration of affective touch. In: Proceedings of the Seventh ACM International Conference on Multimodal Interfaces (ICMI), pp. 222–229 (2005)

Chapter 6
Crowdsourcing Haptic Data Collection

Abstract In this chapter, we investigate the feasibility of large-scale haptic evaluation on online platforms such as Amazon Mechanical Turk. We introduce *proxy modalities* as a way to crowdsource vibrotactile sensations by reliably communicating high-level features through a crowd-accessible channel. We investigate two proxy modalities to represent a high-fidelity tactor: a new vibrotactile visualization, and low-fidelity vibratory translations playable on commodity smartphones. We translated 10 high-fidelity vibrations into both modalities, and in two user studies found that both proxy modalities can communicate affective features, and are consistent when deployed remotely over Mechanical Turk. We analyze fit of features to modalities, and suggest future improvements.

6.1 Introduction

In modern handheld and wearable devices, vibrotactile feedback can provide unintrusive, potentially meaningful cues through wearables in on-the-go contexts [1]. With consumer wearables like Pebble and the Apple Watch featuring high-fidelity actuators, vibrotactile feedback is becoming standard in more user tools. Today, vibrotactile designers seek to provide sensations with various perceptual and emotional connotations to support the growing use cases for vibrotactile feedback (everyday apps, games, etc.). Although low-level design guidelines exist and are helpful for addressing perceptual requirements [2–6], higher-level concerns and design approaches to increase their usability and information capacity (e.g., a user's desired affective response, or affective or metaphorical interpretation) have only recently received study and are far from solved [7–11]. Tactile design thus relies heavily on iteration and user feedback [12]. Despite its importance, collecting user feedback on perceptual and emotional (i.e., affective) properties of tactile sensations in small-scale lab studies is undermined by noise due to individual differences.

© Springer Nature Switzerland AG 2019

H. Seifi, *Personalizing Haptics*, Springer Series on Touch
and Haptic Systems, https://doi.org/10.1007/978-3-030-11379-7_6

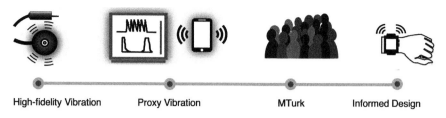

High-fidelity Vibration   Proxy Vibration   MTurk   Informed Design

Fig. 6.1 Conceptual sketch of crowdsourcing data collection for high fidelity vibrations

In other design domains, crowdsourcing enables collecting feedback at scale. Researchers and designers use platforms like Amazon's Mechanical Turk[1] to deploy user studies with large samples, receiving extremely rapid feedback in, e.g., creative text production [13], graphic design [14] and sonic imitations [15].

The problem with crowdsourcing tactile feedback is that the "crowd" can't feel the stimuli. Even when consumer devices have tactors, output quality and intensity is unpredictable and uncontrollable. Sending each user a device is impractical.

What we need are crowd-friendly proxies for test stimuli. Here, we define a *proxy vibration* as a sensation that communicates key characteristics of a source stimulus within a bounded error; a *proxy modality* is the perceptual channel and representation employed. In the new evaluation process thus enabled, the designer translates a sensation of interest into a proxy modality, receives rapid feedback from a crowd-sourcing platform, then interprets that feedback using known error bounds (Fig. 6.1). In this way, designers can receive high-volume, rapid feedback to use in tandem with costly in-lab studies, for example, to guide initial designs or to generalize findings from smaller studies with a larger sample.

To this end, we must first establish feasibility of this approach, with specific goals: **(G1)** Do proxy modalities work? Can they effectively communicate both physical vibrotactile properties (e.g., duration), and high-level affective properties (roughness, pleasantness)? **(G2)** Can proxies be deployed remotely? **(G3)** What modalities work, and **(G4)** what obstacles must be overcome to make this approach practical?

This chapter describes a proof-of-concept for proxy modalities for tactile crowd-sourcing, and identifies challenges throughout the workflow pipeline. We describe and assess two modalities' development, translation process, validation with a test set translation, and MTurk deployment. Our two modalities are a new technique to graphically visualize high-level traits, and the low-fidelity actuators on users' own commodity smartphones. Our test material is a set of 10 vibrotactile stimuli designed for a high-fidelity tactile display suitable for wearables (referred to as "high fidelity vibrations"), and perceptually well understood as presented by that type of display (Fig. 6.6). We conducted two coupled studies, first validating proxy expressiveness in lab, then establishing correspondence of results in remote deployment. Our contributions are:

---

[1] https://www.mturk.com/.

- A way to crowdsource tactile sensations (vibration proxies), with a technical proof-of-concept.
- A visualization method that communicates high-level affective features more effectively than the current tactile visualization standard (vibration waveforms).
- Evidence that both proxy modalities can represent high-level affective features, with lessons about which features work best with which modalities.
- Evidence that our proxy modalities are consistently rated in-lab and remotely, with initial lessons for compliance.

6.2 Related Work

We cover work related to vibrotactile icons and evaluation methods for vibrotactile effects, the current understanding of affective haptics, and work with Mechanical Turk in other modalities.

6.2.1 Existing Evaluation Methods for Vibrotactile Effects

The haptic community has appropriated or developed many types of user studies to evaluate vibrotactile effects and support vibrotactile design. These target a variety of objectives:

(1) *Perceptibility*: Determine the perceptual threshold or Just Noticeable Difference (JND) of vibrotactile parameters. Researchers vary the values of a vibrotactile parameter (e.g., frequency) to determine the minimum perceptible change [16, 17].

(2) *Illusions*: Studies investigate effects like masking or apparent motion of vibrotactile sensations, useful to expand a haptics designer's palette [18–20].

(3) *Perceptual organization*: Reveal the underlying dimensionality of how humans perceive vibrotactile effects (which are generally different than the machine parameters used to generate the stimuli). Multidimensional Scaling (MDS) studies are common, inviting participants compare or group vibrations based on perceived similarity [4, 21–24].

(4) *Encoding abstract information*: Researchers examine salient and memorable vibrotactile parameters (e.g. energy, rhythm) as well as the number of vibrotactile icons that people can remember and attribute to an information piece [4, 6, 24, 25].

(5) *Assign affect*: Studies investigate the link between affective characteristics of vibrations (e.g., pleasantness, urgency) to their engineering parameters (e.g., frequency, waveform) [4, 26–28]. To achieve this, vibrotactile researchers commonly design or collect a set of vibrations and ask participants to rate them on a set of qualitative metrics.

(6) *Identify language*: Participants describe or annotate tactile stimuli in natural language [4, 7, 24, 29, 30].

(7) *Use case support*: Case studies focus on conveying information with vibro-tactile icons such as collaboration [24], public transit [1] and direction [1, 8], or timing of a presentation [31]. In other cases, vibrotactile effects are designed for user engagement, for example in games and movies, multimodal storytelling, or art installations [9, 32]. Here, the designers use iterative design and user feedback (qualitative and quantitative with user rating) to refine and ensure effective design.

All of the above studies would benefit from the large number of participants and fast data collection on MTurk. In this chapter, we chose our methodology so that the results are informative for a broad range of these studies.

6.2.2 Affective Haptics

Vibrotactile designers have the challenge of creating perceptually salient icon sets that convey meaningful content. A full range of expressiveness means manipulating not only a vibration's physical characteristics but also its perceptual and emotional properties, and collecting feedback on this. Here, we refer to all these properties as affective characteristics.

Some foundations for affective vibrotactile design are in place. Studies on tactile language and affect are establishing a set of perceptual metrics [7, 29]. Guest et al. collated a large list of emotion and sensation words describing tactile stimuli; then, based on multidimensional scaling of similarity ratings, proposed *comfort* or *pleasantness* and *arousal* as key dimensions for tactile emotion words, and *rough/smooth*, *cold/warm*, and *wet/dry* for sensation [29]. Even so, there is not yet agreement on an affective tactile design language [10].

In Chap. 4, we compiled research on tactile language into five taxonomies for describing vibrations. (1) *Physical properties* that can be measured: e.g., duration, energy, tempo or speed, rhythm structure; (2) *sensory properties*: roughness, and sensory words from Guest et al.'s touch dictionary [29]; (3) *emotional interpretations*: pleasantness, arousal (urgency), dictionary emotion words [29]; (4) *metaphors* provide familiar examples resembling the vibration's feel: heartbeat, insects; (5) *usage examples* describe events which a vibration fits: an incoming message or alarm.

To evaluate our vibration proxies, we derived six metrics from these taxonomies to capture vibrations' physical, sensory and emotional aspects: (1) duration, (2) energy, (3) speed, (4) roughness, (5) pleasantness, and (6) urgency.

6.2.3 Mechanical Turk (MTurk)

MTurk is a platform for receiving feedback from a large number of users, in a short time at a low cost [33, 34]. These large, fast, cheap samples have proved useful for many cases including running perceptual studies [34], developing taxonomies [35], feedback on text [13], graphic design [14], and sonic imitations [15].

Crowdsourced studies have drawbacks. The remote, asynchronous study environment is not controlled; compared to a quiet lab, participants may be subjected to unknown interruptions, and may spend less time on task with more response variability [33]. MTurk is not suitable for collecting rich, qualitative feedback or following up on performance or strategy [36]. Best practices—e.g., simplifying tasks to be confined to a singular activity, or using instructions complemented with example responses—are used to reduce task ambiguity and improve response quality [37]. Some participants try to exploit the service for personal profit, exhibiting low task engagement [38], and must be pre- or post-screened.

Studies have examined MTurk result validity in other domains. Most relevantly, Heer et al. [34] validated MTurk data for graphical perception experiments (spatial encoding and luminance contrast) by replicating previous perceptual studies on MTurk. Similarly, we compare results of our local user study with an MTurk study to assess viability of running vibrotactile studies on MTurk, and collect and examine phone properties in our MTurk deployment.

Need for HapTurk: Our present goal is to give the haptic design community access to crowdsourced evaluation so we can establish modality-specific methodological tradeoffs. There is ample need for huge-sample haptic evaluation. User experience of transmitted sensations must be robust to receiving device diversity. Techniques to broadcast haptic effects to video [39, 40], e.g., with YouTube [41] or MPEG7 [42, 43] now require known high-fidelity devices because of remote device uncertainty; the same applies to social protocols developed for remote use of high-quality vibrations, e.g. in collaborative turn taking [24]. Elsewhere, studies of vibrotactile use in consumer devices need larger samples: e.g., perceivability [44], encoding of caller parameters [45], including caller emotion and physical presence collected from pressure on another handset [46], and usability of expressive, customizable vibrotactile icons in social messaging [47]. To our knowledge, this is the first attempt to run a haptic study on a crowdsource site and characterize its feasibility and challenges for haptics.

6.3 Sourcing Reference Vibrations and Qualities

We required a set of exemplar source vibrations on which to base our proxy modalities. This set needed to (1) vary in physical, perceptual, and emotional characteristics, (2) represent the variation in a larger source library, and (3) be small enough for experimental feasibility.

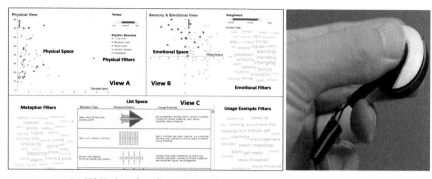

(a) *VibViz* interface from chapter 4   (b) C2 tactor

Fig. 6.2 Source of high-fidelity vibrations and perceptual rating scales

6.3.1 High-Fidelity Reference Library

We chose 10 vibrations from a large, freely available library of 120 vibrations (*VibViz*, Chap. 4), browsable through five descriptive facets,[2] and ratings of facet properties. Vibrations were designed for an Engineering Acoustics C2 tactor, a high-fidelity, wearable-suitable voice coil, commonly used in haptic research. We employed *VibViz*'s filtering tools to sample, ensuring variety and coverage by selecting vibrations at high and low ends of energy/duration dimensions, and filtering by ratings of temporal structure/rhythm, roughness, pleasantness, and urgency. To reduce bias, two researchers independently and iteratively selected a set of 10 items each, which were then merged.

Because *VibViz* was designed for a C2 tactor, we used a handheld C2 in the present study (Fig. 6.2b).

6.3.2 Affective Properties and Rating Scales

To evaluate our proxies, we adapted six rating scales from the tactile literature and new studies. In Chap. 4, we proposed five facets for describing vibrations including physical, sensory, emotional, metaphors, and use examples. Three facets comprise quantitative metrics and adjectives; two use descriptive words.

We chose six quantitative metrics from Chap. 4 that capture important affective (physical, perceptual, and emotional) vibrotactile qualities: (1) *duration* [low-high], (2) *energy* [low-high], (3) *speed* [slow-fast], (4) *roughness* [smooth-rough], (5) *urgency* [relaxed-alarming], and (6) *pleasantness* [unpleasant-pleasant]. A large scale (0–100) allowed us to treat the ratings as continuous variables. To keep trials quick and MTurk-suitable, we did not request open-ended responses or tagging.

---

[2]Called taxonomy in the original conference publication.

6.4 Proxy Choice and Design

The proxies' purpose was to capture high-level traits of source signals. We investigated two proxy channels and approaches, to efficiently establish viability and search for triangulated perspectives on what will work. The most obvious starting points are to (1) visually augment the current standard of a direct trace of *amplitude* $= f$ (*time*), and (2) reconstruct vibrations for common-denominator, low-fidelity actuators.

We considered other possibilities (e.g., auditory stimuli, for which MTurk has been used [15], or animations). However, our selected modalities balance a) directness of translation (low fidelity could not be excluded); b) signal control (hard to ensure consistent audio quality/volume/ambient masking); and c) development progression (visualization underlies animation, and is simpler to design, implement, display). We avoided multisensory combinations at this early stage for clarity of results. Once the key modalities are tested, combinations can be investigated in future work.

"Ref" denotes high-fidelity source renderings (C2 tactor).

(1) **Visual proxies**: Norms in published works (e.g., [24]) directed our work in Chap. 4 to confirm that users rely on graphical f (*time*) plots to skim and choose from large libraries. We tested the direct plot, Vis_{dir}, as the status-quo representation.

However, these unmodified time-series emphasize or mask traits differently than felt vibrations, in particular for higher-level or "meta" responses. We considered many other means of visualizing vibration characteristics, pruned candidates and refined design via piloting to produce a new scheme which explicitly *emphasizes* affective features, Vis_{emph}.

(2) **Low-fidelity vibration proxy**: Commodity device (e.g., smartphone) actuators usually have low output capability compared to the C2, in terms of frequency response, loudness range, distortion and parameter independence. Encouraged by expressive rendering of vibrotactile sensations with commodity actuation (from early constraints [24] to deliberate design-for-lofi [47]), we altered stimuli to convey high-level parameters under these conditions, hereafter referred to as LofiVib.

Translation: Below, we detail first-pass proxy development. In this feasibility stage, we translated proxy vibrations manually and iteratively, as we sought generalizable mappings of the parametric vibration definition to the perceptual quality we wished to highlight in the proxy. We frequently relied on a cycle of user feedback, e.g., to establish the perceived roughness of the original stimuli and proxy candidate.

Automatic translation is an exciting goal. Without it, HapTurk is still useful for gathering large samples; but automation will enable a very rapid create-test cycle. It should be attainable, bootstrapped by the up-scaling of crowdsourcing itself. With a basic process in place, we can use MTurk studies to identify these mappings relatively quickly.

Fig. 6.3 Vis$_{dir}$ Visualization, based on VibViz

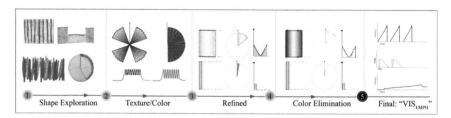

Fig. 6.4 Visualization design process. Iterative development and piloting results in the Vis$_{emph}$ visualization pattern

6.4.1   Visualization Design (Vis$_{dir}$ and Vis$_{emph}$)

Vis$_{dir}$ was based on the original waveform visualization used in *VibViz* (Fig. 6.3). In Matlab, vibration frequency and envelope were encoded to highlight its pattern over time. Since Vis$_{dir}$ patterns were detailed, technical and often inscrutable for users without an engineering background, we also developed a more interpretive visual representation, Vis$_{emph}$; and included Vis$_{dir}$ as a status-quo baseline.

We took many approaches to depicting vibration high-level properties, with visual elements such as line thickness, shape, texture and colour (Fig. 6.4). We first focused on line sharpness, colour intensity, length and texture: graphical waveform smoothness and roughness were mapped to perceived roughness; colour intensity highlighted perceived energy. Duration mapped to length of the graphic, while colour and texture encoded the original's invoked emotion. Four participants were informally interviewed and asked to feel Ref vibrations, describe their reactions, and compare them to several visualization candidates. Participants differed in their responses, and had difficulties in understanding vibrotactile emotional characteristics from the graphic (i.e., pleasantness, urgency), and in reading the circular patterns. We simplified the designs, eliminating representation of emotional characteristics (color, texture), while retaining more objective mappings for physical and sensory characteristics.

Vis$_{emph}$ won an informal evaluation of final proxy candidates ($n = 7$), and was captured in a translation guideline (Figs. 6.5 and 6.6).

Example	Roughness	Energy		Duration
	by the line's **roughness**	by the line's **thickness** &	by **height**	by the **length** of the x-axis
	rough ~~~~~~~ so-so ———— smooth ————	high ———— medium ——— low ————	high medium low	longest ———— short ———|— — ⊣ (compared to the longest)

Fig. 6.5 Final Vis$_{emph}$ visualization guide, used by researchers to create Vis$_{emph}$ proxy vibrations and provided to participants during Vis$_{emph}$ study conditions

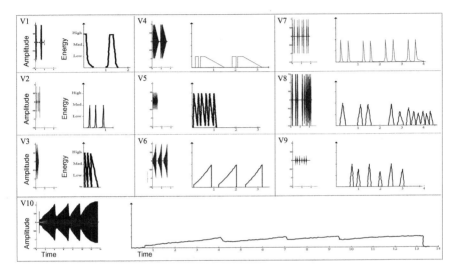

Fig. 6.6 Vibrations visualized as both Vis$_{dir}$ (left of each pair) and Vis$_{emph}$

6.4.2 Low Fidelity Vibration Design

For our second proxy modality, we translated Ref vibrations into LofiVib vibrations. We used a smartphone platform for their built-in commodity-level vibrotactile displays, their ubiquity amongst users, and low security concerns for vibration imports to personal devices [48]. To distribute vibrations remotely, we used HTML5 Vibration API, implemented on Android phones running compatible web browsers (Google Chrome or Mozilla Firefox).

As with Vis$_{emph}$, we focused on physical properties when developing LofiVib (our single low-fi proxy exemplar). We emphasized rhythm structure, an important design parameter [4] and the only direct control parameter of the HTML5 API, which issues vibrations using a series of on/off durations. Simultaneously, we manipulated perceived energy level by adjusting the actuator pulse train on/off ratio, up to the point where the rhythm presentation was compromised. Shorter durations represented a weak-feeling hi-fi signal, while longer durations conveyed intensity in the original.

Fig. 6.7 Example of
LofiVib proxy design. Pulse
duration was hand-tuned to
represent length and
intensity, using duty cycle to
express dynamics such as
ramps and oscillations

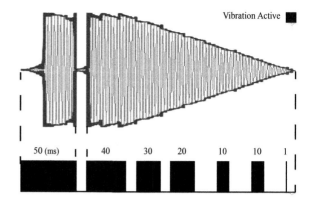

This was most challenging for dynamic intensities or frequencies, such as increasing or decreasing ramps, and long, low-intensity sensations. Here we used a duty-cycle inspired technique, similar to [47], illustrated in Fig. 6.7.

To mitigate the effect of different actuators found in smartphones, we limited our investigation to Android OS. While this restricted our participant pool, there was nevertheless no difficulty in quickly collecting data for either study. We designed for two phones representing the largest classes of smartphone actuators: Samsung Galaxy Nexus, which contains a coin-style actuator, and a Sony Xperia Z3 Compact, which uses a pager motor resulting in more subdued, smooth sensations. Though perceptually different, control of both actuator styles are limited to on/off durations. As with Vis$_{emph}$, we developed LofiVib vibrations iteratively, first with team feedback, then informal interviews (n = 6).

6.5   Study 1: In-Lab Proxy Vibration Validation (G1)

We obtained user ratings for the hi-fi source vibrations Ref and three proxies (Vis$_{dir}$, Vis$_{emph}$, and LofiVib). An in-lab format avoided confounds and unknowns due to remote MTurk deployment, addressed in Study 2. Study 1 had two versions: in one, participants rated visual proxies Vis$_{dir}$ and Vis$_{emph}$ next to Ref; and in the other, LofiVib next to Ref. Ref$_{Vis}$ and Ref$_{LofiVib}$ denote these two references, each compared with its respective proxy(ies) and thus with its own data. In each substudy, participants rated each Ref vibration on 6 scales [0–100] in a computer survey, and again for the proxies. Participants in the visual substudy did this for both Vis$_{dir}$ and Vis$_{emph}$, then indicated preference for one. Participants in the lo-fi study completed the LofiVib survey on a phone, which also played vibrations using Javascript and HTML5; other survey elements employed a laptop. 40 participants aged 18–50 were recruited via university undergraduate mailing lists. 20 (8F) participated in the visual substudy, and a different 20 (10F) in the low-fi vibration substudy.

Reference and proxies were presented in different random orders. Pilots confirmed that participants did not notice proxy/target linkages, and thus were unlikely to consciously match their ratings between pair elements. Ref/proxy presentation order was counterbalanced, as was $\mathrm{Vis}_{dir}/\mathrm{Vis}_{emph}$.

6.5.1 Comparison Metric: Equivalence Threshold

To assess whether proxy modalities were rated similarly to their targets, we employed *equivalence testing*, which tests the hypothesis that sample means are within a threshold δ, against the null of being outside it [49]. This tests if two samples are equivalent with a known error bound; it corresponds to creating confidence intervals of means, and examining whether they lie entirely within the range $(-\delta, \delta)$.

We first computed least-squares means for the 6 rating scales for each proxy modality and vibration. 95% confidence intervals (CI) for Ref rating means ranged from 14.23 points (Duration ratings) to 20.33 (Speed). Because estimates of the Ref "gold standard" mean could not be more precise than these bounds, we set equivalence thresholds for each rating equal to CI width. For example, given the CI for Duration of 14.23, we considered proxy Duration ratings equivalent if the CI for a difference fell completely in the range $(-14.23, 14.23)$. With pooled standard error, this corresponded to the case where two CIs overlap by more than 50%. We also report when a *difference* was detected, through typical hypothesis testing (i.e., where CIs do not overlap).

Thus, each rating set pair could be *equivalent*, *uncertain*, or *different*. Figure 6.8 offers insight into how these levels are reflected in the data given the high rating variance. This approach gives a useful error bound, quantifying the precision tradeoff in using vibration proxies to crowdsource feedback.

6.5.2 Proxy Validation (Study 1) Results and Discussion

6.5.2.1 Overview of Results

Study 1 results appear graphically in Fig. 6.9. To interpret this plot, look for (1) equivalence indicated by bar color, and CI size by bar height (dark green/small are good); (2) rating richness: how much spread, vibration to vibration, within a cell indicates how well that parameter captures the differences users perceived; (3) modality consistency: the degree to which the bars' up/down pattern translates vertically across rows. When similar (and not flat), the proxy translations are being interpreted by users in the same way, providing another level of validation. We structure our discussion around how the three modalities represent the different rating scales. We refer to the number of *equivalents* and *differents* in a given cell as [x:z], with y = number of *uncertains*, and $x + y + z = 10$.

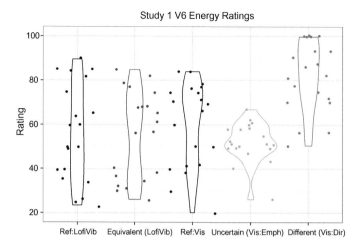

Fig. 6.8 Rating distributions from Study 1, using V6 Energy as an example. These violin plots illustrate (1) the large variance in participant ratings, and (2) how equivalence thresholds reflect the data. When equivalent, proxy ratings are visibly similar to Ref. When uncertain, ratings follow a distribution with unclear differences. When different, there is a clear shift

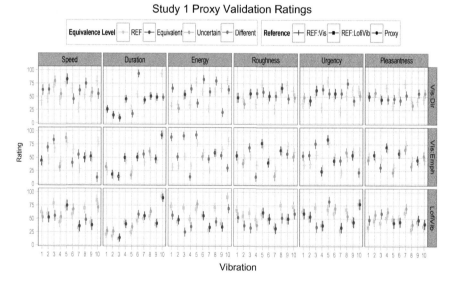

Fig. 6.9 95% confidence intervals and equivalence test results for Study 1—Proxy Validation. Grey represents Ref ratings. Dark green maps equivalence within our defined threshold, and red a statistical difference indicating an introduced bias; light green results are inconclusive. Within each cell, variation of Ref ratings means vibrations were rated differently compared to each other, suggesting they have different perceptual features and represent a varied set of source stimuli

6.5.2.2 Duration and Pleasantness Were Translatable

Duration was comparably translatable for LofiVib [5:1] and Vis_{emph} [6:1]; Vis_{dir} was less consistent [7:3] (two differences very large). Between the three modalities, 9/10 vibrations achieved equivalence with at least one modality.

For Duration, this is unsurprising. It is a physical property that is controllable through the Android vibration API, and both visualization methods explicitly present Duration as their x-axis. This information was apparently not lost in translation.

More surprisingly, Pleasantness fared only slightly worse for LofiVib [4:2] and Vis_{emph} [4:1]; 8/10 vibrations had at least one modality that provided equivalence. Pleasantness is a higher-level affective feature than Duration. Although not an absolute victory, this result gives evidence that, with improvement, crowdsourcing may be a viable method of feedback for at least one affective parameter.

6.5.2.3 Speed and Urgency Translated Better with LofiVib

LofiVib was effective at representing Urgency [6:2]; Vis_{emph} attained only [4:5], and Vis_{dir} [3:5]. Speed was less translatable. LofiVib did best at [4:2]; Vis_{dir} reached only [1:6], and Vis_{emph} [3:5]. However, the modalities again complemented each other. Of the three, 9/10 vibrations were equivalent at least once for Urgency (V8 was not). Speed had less coverage: 6/10 had equivalencies (V3, 4, 6, 10 did not).

6.5.2.4 Roughness Had Mixed Results; Best with Vis_{emph}

Roughness ratings varied heavily by vibration. 7 vibrations had at least one equivalence (V2, 4, 10 did not). All modalities had 4 equivalencies each: Vis_{emph} [4:3], Vis_{dir} [4:4], and LofiVib [4:5].

6.5.2.5 Energy Was Most Challenging

Like Roughness, 7 vibrations had at least one equivalence between modalities (V1, 4, 10 did not). LofiVib [4:5] did best with Energy; Vis_{emph} and Vis_{dir} struggled at [1:8].

6.5.2.6 Emphasized Visualization Outperformed Direct Plot

Though it depended on the vibration, Vis_{emph} outperformed Vis_{dir} for most metrics, having the same or better equivalencies/differences for Speed, Energy, Roughness, Urgency, and Pleasantness. Duration was the only mixed result, as Vis_{dir} had both more equivalencies and more differences [7:3] versus [6:1]. In addition, 16/20 participants (80%) preferred Vis_{emph} to Vis_{dir}. Although not always clear-cut, these comparisons overall indicate that our Vis$_{emph}$ visualization method communicated these affective qualities more effectively than the status quo. This supports our approach to emphasized visualization, and motivates the future pursuit of other visualizations.

6.5.2.7   V4, V10 Difficult, V9 Easy to Translate

While most vibrations had at least one equivalency for 5 rating scales, V4 and V10 only had 3. V4 and V10 had no equivalences at all for Speed, Roughness, and Energy, making them some of the most difficult vibrations to translate. V4's visualization had very straight lines, perhaps downplaying its texture. V10 was by far the longest vibration, at 13.5 s (next longest was V8 with 4.4 s). Its length may have similarly masked textural features.

V8 was not found to be equivalent for Urgency and Pleasantness. V8 is an extremely irregular vibration, with a varied rhythm and amplitude, and the second longest. This may have made it difficult to glean more intentional qualities like Urgency and Pleasantness. However, it was only found to be different for Vis$_{dir}$/Urgency, so we cannot conclude that significant biases exist.

By contrast, V9 was the only vibration that had an equivalency for every rating scale, and in fact could be represented across all ratings with LofiVib. V9 was a set of distinct pulses, with no dynamic ramps; it thus may have been well suited to translation to LofiVib.

6.5.2.8   Summary

In general, these results indicate promise, but also need improvement and combination of proxy modalities. Unsurprisingly, participant ratings varied, reducing confidence and increasing the width of confidence intervals (indeed, this is partial motivation to access larger samples). Even so, both differences and equivalencies were found in every rating/proxy modality pairing. Most vibrations were equivalent with at least one modality, suggesting that we might pick an appropriate proxy modality depending on the vibration; we discuss the idea of triangulation in more detail later. Duration and Pleasantness were fairly well represented, Urgency and Speed were captured best by LofiVib, and Roughness was mixed. Energy was particularly difficult to represent with these modalities. We also find that results varied depending on vibration, meaning that more analysis into what makes vibrations easier or more difficult to represent could be helpful.

Though we were able to represent several features using proxy modalities within a bounded error rate, this alone does not mean they are crowdsource-friendly. All results from Study 1 were gathered in-lab, a more controlled environment than over MTurk. We thus ran a second study to validate our proxy modality ratings when deployed remotely.

6.6 Study 2: Deployment Validation with MTurk (G2)

To determine whether rating of a proxy is similar when gathered locally or remotely, we deployed the same computer-run proxy modality surveys on MTurk. We wanted to discover the challenges all through the pipeline for running a vibrotactile study on MTurk, including larger variations in phone actuators and experimental conditions (G4). We purposefully did not iterate on our proxy vibrations or survey, despite identifying many ways to improve them, to avoid creating a confound in comparing results of the two studies.

The visualization proxies were run as a single MTurk Human Intelligence Task (HIT), counterbalanced for order; the LofiVib survey was deployed as its own HIT. Each HIT was estimated at 30 m, for which participants received $2.25 USD. In comparison, Study 1 participants were estimated to take 1 h and received $10 CAD. We anticipated a discrepancy in average task time due to a lack of direct supervision for the MTurk participants, and expected this to lead to less accurate participant responses, prompting the lower payrate. On average, it took 7 m for participants to complete the HIT while local study participants took 30 m.

We initially accepted participants of any HIT approval rate to maximize recruitment in a short timeframe. Participants were post-screened to prevent participation in both studies. 49 participants were recruited. No post-screening was used for the visual sub-study. For the LofiVib proxy survey, we post-screened to verify device used [36]. We asked participants (a) confirm their study completion with an Android device via a survey question, (b) detected actual device via FluidSurvey's OS-check feature, and (c) rejected inconsistent samples (e.g., 9 used non-Android platforms for LofiVib). Of the included data, 20 participants participated each in the visual proxy condition (6F) and the LofiVib condition (9F).

For both studies, Study 1's data was used as a "gold standard" that served as a baseline comparison with the more reliable local participant ratings [37]. We compared the remote proxy results (from MTurk) to the Ref results gathered in Study 1, using the same analysis methods.

6.6.1 Results

Study 2 results appear in Fig. 6.10, which compares remotely collected ratings with locally collected ratings for the respective reference (the same reference as for Fig. 6.9). It can be read the same way, but adds information. Based an analysis of a different comparison, a red star indicates a statistically significant difference between remote proxy ratings and corresponding local *proxy* ratings. This analysis revealed that ratings for the same proxy gathered remotely and locally disagreed 21 times (stars) out of 180 rating/modality/vibration combination; i.e., relatively infrequently.

Overall, we found similar results and patterns in Study 2 as for Study 1. The two figures show similar up/down rating patterns; the occasional exceptions correspond

Study 2 MTurk Deployment Ratings

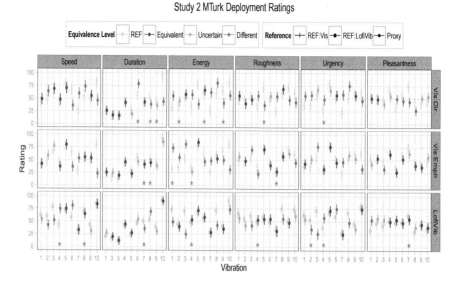

Fig. 6.10 95% Confidence Intervals and Equivalence Test Results for Study 2—MTurk Deployment Validation. Equivalence is indicated with dark green, difference is indicated with red, and uncertainty with light green. Red star indicates statistically significant difference between remote and local proxy ratings

to red-starred items. Specific results varied, possibly due to statistical noise and rating variance. We draw similar conclusions: that proxy modalities can still be viable when deployed on MTurk, but require further development to be reliable in some cases.

6.7 Discussion

Here, we discuss high level implications from our findings and relate them to our study goals (G1-G4 in Introduction).

6.7.1 Proxy Modalities Are Viable for Crowdsourcing (G1, G2: Feasibility)

Our studies showed that proxy modalities can represent affective qualities of vibrations within reasonably chosen error bounds, depending on the vibration. These results largely translate to deployment on MTurk. Together, these two steps indicate that proxy modalities are a viable approach to crowdsourcing vibrotactile sensations, and can reach a usable state with a bounded design iteration (as outlined in the following sections). This evidence also suggests that we may be able to deploy directly to MTurk for future validation. Our two-step validation was important as a first look at whether ratings shift dramatically; and we saw no indications of bias or overall shift between locally running proxy modalities and remotely deploying them.

6.7.2 Triangulation (G3: Promising Directions/Proxies)

Most vibrations received equivalent ratings for most scales in at least one proxy modality. Using proxy modalities in tandem might help improve response accuracy. For example, V6 could be rendered with LofiVib for a pleasantness rating, then as Vis_{emph} for Urgency. Alternatively, we might develop an improved proxy vibration by combining modalities—a visualization with an accompanying low-fidelity vibration.

6.7.3 Animate Visualizations (G3: Promising Directions)

Speed and Urgency were not as effectively transmitted with our visualizations as with our vibration. Nor was Duration well portrayed with Vis_{dir}, which had a shorter time axis than the exaggerated Vis_{emph}. It may be more difficult for visual representations to portray time effectively: perhaps it is hard for users to distinguish Speed/Urgency, or the time axis is not at an effective granularity. Animations (e.g., adding a moving line to help indicate speed and urgency), might help to decouple these features. As with triangulation, this might also be accomplished through multimodal proxies which augment a visualization with a time-varying sense using sounds or vibration. Note, however, that Duration was more accurately portrayed by Vis_{emph}, suggesting that direct representation of physical features *can* be translated.

6.7.4 Sound Could Represent Energy (G3: Promising Directions)

Our high-fidelity reference is a voice-coil actuator, also used in audio applications. Indeed, in initial pilots we played vibration sound files through speakers. Sound is the closest to vibration in the literature, and a vibration signal's sound output is correlated with the vibration energy and sensation.

However, in our pilots, sometimes the vibration sound did not match the sensation; was not audible (low frequency vibrations); or the C2 could only play part of the sound (i.e, the sound was louder than the sensation).

Thus, while the raw sound files are not directly translatable, a sound proxy definitely has potential. It could, for example, supplement where the Vis_{dir} waveform failed to perform well on any metric (aside from Duration) but a more expressive visual proxy (Vis_{emph}) performed better.

6.7.5   Device Dependency and Need for Energy Model for Vibrations (G4: Challenges)

Energy did not translate well. This could be a linguistic confusion, but also a failure to translate this feature. For the visualization proxies, it may be a matter of finding the right representation, which we continue to work on.

However, with LofiVib, this represents a more fundamental tradeoff due to characteristics of phone actuators, which have less control over energy output than we do with a dedicated and more powerful C2 tactor. The highest vibration energy available in phones is lower than for the C2; this additional power obviously extends expressive range. Furthermore, vibration energy and time are coupled in phone actuators: the less time the actuator is on, the lower the vibration energy. As a result, it is difficult to have a very short pulses with very high energy (V1, V3, V8). The C2's voice coil technology does not have this duty-cycle derived coupling. Finally, the granularity of the energy dimension is coarser for phone actuators. This results in a tradeoff for designing (for example) a ramp sensation: if you aim for accurate timing, the resulting vibration would have a lower energy (V10). If you match the energy, the vibration will be longer.

Knowing these tradeoffs, designers and researchers can adjust their designs to obtain more accurate results on their intended metric. Perhaps multiple LofiVib translations can be developed which maintain different qualities (one optimized on timing and rhythm, the other on energy). In both these cases, accurate models for rendering these features will be essential.

6.7.6   Vibrotactile Affective Ratings Are Generally Noisy (G4: Challenges)

Taken as a group, participants were not highly consistent among one another when rating these affective studies, whether local or remote. This is in line with our previous work (Chap. 4), and highlights a need to further develop rating scales for affective touch. Larger sample sizes, perhaps gathered through crowdsourcing, may help reduce or characterize this error. Alternatively, it gives support to the need to develop mechanisms for individual customization. If there are "types" of users who do share preferences and interpretations, crowdsourcing can help with this as well.

6.7.7 Response and Data Quality for MTurk LofiVib Vibrations (G4: Challenges)

When deploying vibrations over MTurk, 8/29 participants (approximately 31%) completed the survey using non-Android based OSes (Mac OS X, Windows 7, 8.1, NT) despite these requirements being listed in the HIT and the survey. One participant reported not being able to feel the vibrations despite using an Android phone. This suggests that enforcing a remote survey to be taken on the phone is challenging, and that additional screens are needed to identify participants not on a particular platform. Future work might investigate additional diagnostic tools to ensure that vibrations are being generated, through programmatic screening of platforms, well-worded questions and instructions, and (possibly) ways of detecting vibrations actually being played, perhaps through the microphone or accelerometer).

6.7.8 Automatic Translation (G4: Challenges)

Our proxy vibrations were developed by hand, to focus on the feasibility of crowdsourcing. However, this additional effort poses a barrier for designers that might negate the benefits of using a platform of MTurk. As this approach becomes better defined, we anticipate automatic translation heuristics for proxy vibrations using validated algorithms. Although these might be challenging to develop for emotional features, physical properties like amplitude, frequency, or measures of energy and roughness would be a suitable first step. Indeed, crowdsourcing itself could be used to create these algorithms, as several candidates could be developed, their proxy vibrations deployed on MTurk, and the most promising algorithms later validated in lab.

6.7.9 Limitations

A potential confound was introduced by Vis_{emph} having a longer time axis than Vis_{dir}: some of Vis_{emph}'s improvements could be due to seeing temporal features in higher resolution. This is exacerbated by V10 being notably longer than the next longest vibration, V8 (13.5 s vs. 4.4 s), further reducing temporal resolution vibrations other than V10.

We presented ratings to participants by-vibration rather than by-rating. Because participants generated all ratings for a single vibration at the same time, it is possible there are correlations between the different metrics. We chose this arrangement because piloting suggested it was less cognitively demanding than presenting metrics separately for each vibration. Future work can help decide whether correlations exist between metrics, and whether these are an artifact of stimulus presentation or an underlying aspect of the touch aesthetic.

Despite MTurk's ability to recruit more participants, we used the same sample size of 40 across both studies. While our proxies seemed viable for remote deployment, there were many unknown factors in MTurk user behaviour at the time of deployment. We could not justify more effort without experiencing these factors firsthand. Thus, we decided to use a minimal sample size for the MTurk study that was statistically comparable to the local studies. In order to justify a larger remote sample size in the future, we believe it is best to iterate the rating scales and to test different sets of candidate modalities.

As discussed, we investigated two proxy modalities in this first examination but look forward to examining others (sound, text, or video) alone or in combination.

6.8 Conclusion

In this chapter, we crowdsourced high-level parameter feedback on vibrotactile sensations using a new method of *proxy vibrations*. We translated our initial set of high-fidelity vibrations, suitable for wearables or other haptic interactions, into two proxy modalities: a new vibrotactile visualization method, and low-fidelity vibrations on phones.

We established the most high-risk aspects of vibrotactile proxies, namely feasibility in conveying affective properties, and consistent local and remote deployment with two user studies. Finally, we highlighted promising directions and challenges of vibrotactile proxies, to guide future tactile crowdsourcing developments, targeted to empower vibrotactile designers with the benefits crowdsourcing brings.

References

1. Brunet, L., Megard, C., Paneels, S., Changeon, G., Lozada, J., Daniel, M.P., Darses, F.: Invitation to the voyage: the design of tactile metaphors to fulfill occasional travelers' needs in transportation networks. In: IEEE World Haptics Conference (WHC '13), pp. 259–264 (2013). https://doi.org/10.1109/WHC.2013.6548418
2. MacLean, K., Enriquez, M.: Perceptual design of haptic icons. In: Proceedings of EuroHaptics Conference, pp. 351–363 (2003)
3. Hwang, I., Seo, J., Kim, M., Choi, S.: Vibrotactile perceived intensity for mobile devices as a function of direction, amplitude, and frequency. IEEE Trans. Haptics (ToH) **6**(3), 352–362 (2013). https://doi.org/10.1109/TOH.2013.2, http://ieeexplore.ieee.org/lpdocs/epic03/wrapper.htm?arnumber=6420831
4. Ternes, D., Maclean, K.E.: Designing large sets of haptic icons with rhythm. In: Haptics: Perception, Devices and Scenarios, pp. 199–208. Springer, Berlin (2008)
5. Brewster, S., Brown, L.M.: Tactons: structured tactile messages for non-visual information display. In: Proceedings of the Fifth Conference on Australasian User Interface, vol.28, pp. 15–23. Australian Computer Society, Inc. (2004)

6. Brown, L.M., Brewster, S.A., Purchase, H.C.: Multidimensional tactons for non-visual information presentation in mobile devices. In: Proceedings of the 8th Conference on Human-Computer Interaction with Mobile Devices and Services (MobileHCI '06), pp. 231–238. ACM Press, New York (2006). https://doi.org/10.1145/1152215.1152265, http://dl.acm.org/citation.cfm?id=1152215.1152265

7. Obrist, M., Seah, S.A., Subramanian, S.: Talking about tactile experiences. In: Proceedings of the ACM SIGCHI Conference on Human Factors in Computing Systems (CHI '13), pp. 1659–1668. ACM (2013)

8. Arab, F., Paneels, S., Anastassova, M., Coeugnet, S., Le Morellec, F., Dommes, A., Chevalier, A.: Haptic patterns and older adults: to repeat or not to repeat? In: Proceedings of IEEE World Haptics Conference (WHC '15), pp. 248–253. IEEE (2015). https://doi.org/10.1109/WHC.2015.7177721, http://ieeexplore.ieee.org/lpdocs/epic03/wrapper.htm?arnumber=7177721

9. Israr, A., Zhao, S., Schwalje, K., Klatzky, R., Lehman, J.: Feel effects: enriching storytelling with haptic feedback. ACM Trans. Appl. Percept. (TAP) 11, 11:1–11:17 (2014)

10. Jansson-Boyd, C.V.: Touch matters: exploring the relationship between consumption and tactile interaction. Soc. Semiot. **21**(4), 531–546 (2011). https://doi.org/10.1080/10350330.2011.591996

11. Okamoto, S., Nagano, H., Yamada, Y.: Psychophysical dimensions of tactile perception of textures. IEEE Trans. Haptics (ToH) **6**(1), 81–93 (2013)

12. Schneider, O.S., MacLean, K.E.: Improvising design with a haptic instrument. In: Proceedings of IEEE Haptics Symposium (HAPTICS '14), pp. 327–332. IEEE (2014)

13. Siangliulue, P., Arnold, K.C., Gajos, K.Z., Dow, S.P.: Toward collaborative ideation at scale - leveraging ideas from others to generate more creative and diverse ideas pao. In: Proceedings of ACM Conference on Computer-Supported Cooperative Work and Social Computing (CSCW '15), pp. 937–945. ACM Press, New York (2015). https://doi.org/10.1145/2675133.2675239, http://dl.acm.org/citation.cfm?id=2675133.2675239

14. Xu, A., Huang, S.W., Bailey, B.: Voyant: Generating structured feedback on visual designs using a crowd of non-experts. In: ACM Conference on Computer-Supported Cooperative Work and Social Computing (CSCW '14), pp. 1433–1444. ACM Press, New York (2014). https://doi.org/10.1145/2531602.2531604, http://dl.acm.org/citation.cfm?id=2531602.2531604

15. Cartwright, M., Pardo, B.: VocalSketch: vocally imitating audio concepts. In: Proceedings of the ACM SIGCHI Conference on Human Factors in Computing Systems (CHI '15), pp. 43–46. ACM Press, New York (2015). https://doi.org/10.1145/2702123.2702387, http://dl.acm.org/citation.cfm?id=2702123.2702387

16. Pongrac, H.: Vibrotactile perception: examining the coding of vibrations and the just noticeable difference under various conditions. Multimed. Syst. **13**(4), 297–307 (2008). https://doi.org/10.1007/s00530-007-0105-x

17. MacLean, K.E.: Foundations of transparency in tactile information design. IEEE Trans. Haptics (ToH) **1**(2), 84–95 (2008)

18. Hayward, V.: A brief taxonomy of tactile illusions and demonstrations that can be done in a hardware store. Brain Res. Bull. **75**(6), 742–752 (2008). https://doi.org/10.1016/j.brainresbull.2008.01.008, http://www.sciencedirect.com/science/article/pii/S0361923008000178

19. Israr, A., Poupyrev, I.: Tactile brush: drawing on skin with a tactile grid display. In: Proceedings of the ACM SIGCHI Conference on Human Factors in Computing Systems (CHI '11), pp. 2019–2028. ACM Press, Vancouver (2011). https://doi.org/10.1145/1978942.1979235, http://dl.acm.org/citation.cfm?id=1978942.1979235

20. Seo, J., Choi, S.: Perceptual analysis of vibrotactile flows on a mobile device. IEEE Trans. Haptics (ToH) **6**(4), 522–527 (2013). https://doi.org/10.1109/TOH.2013.24, http://www.ncbi.nlm.nih.gov/pubmed/24808404

21. Hollins, M., Faldowski, R., Rao, S., Young, F.: Perceptual dimensions of tactile surface texture: a multidimensional scaling analysis. Percept. Psychophys. **54**, 697–705 (1993)

22. van Erp, J.B., Spapé, M.M.: Distilling the underlying dimensions of tactile melodies. Proc. Eurohaptics Conf. **2003**, 111–120 (2003)

23. Pasquero, J., Luk, J., Little, S., Maclean, K.: Perceptual analysis of haptic icons: an investigation into the validity of cluster sorted mds. In: Proceedings of IEEE Haptic Symposium (HAPTICS '06), pp. 437–444. IEEE, Alexandria (2006)
24. Chan, A., MacLean, K.E., McGrenere, J.: Designing haptic icons to support collaborative turn-taking. Int. J. Hum. Comput. Stud. (IJHCS) **66**, 333–355 (2008)
25. Allen, M., Gluck, J., MacLean, K.E., Tang, E.: An initial usability assessment for symbolic haptic rendering of music parameters. In: Proceedings of 7th International Conference on Multimodal Interfaces (ICMI '05), pp. 244–251, Trento, Italy (2005)
26. Yoo, Y., Yoo, T., Kong, J., Choi, S.: Emotional responses of tactile icons: effects of amplitude, frequency, duration, and envelope. In: Proceedings of IEEE World Haptics Conference (WHC'15), pp. 235–240 (2015). https://doi.org/10.1109/WHC.2015.7177719
27. Raisamo, J., Raisamo, R., Surakka, V.: Comparison of saltation, amplitude modulation, and a hybrid method of vibrotactile stimulation. IEEE Trans. Haptics (ToH) **6**(4), 517–521 (2013). https://doi.org/10.1109/TOH.2013.25
28. Koskinen, E., Kaaresoja, T., Laitinen, P.: Feel-good touch: finding the most pleasant tactile feedback for a mobile touch screen button. In: Proceedings of the 10th International Conference on Multimodal Interfaces (ICMI '08), pp. 297–304. ACM, New York (2008). https://doi.org/10.1145/1452392.1452453, http://doi.acm.org.ezproxy.library.ubc.ca/10.1145/1452392.1452453
29. Guest, S., Dessirier, J.M., Mehrabyan, A., McGlone, F., Essick, G., Gescheider, G., Fontana, A., Xiong, R., Ackerley, R., Blot, K.: The development and validation of sensory and emotional scales of touch perception. Atten. Percept. Psychophys. **73**(2), 531–550 (2011)
30. Hwang, I., MacLean, K.E., Brehmer, M., Hendy, J., Sotirakopoulos, A., Choi, S.: The haptic crayola effect: exploring the role of naming in learning haptic stimuli. In: Proceedings of IEE World Haptics Conference (WHC '11), pp. 385–390. IEEE, Istanbul (2011)
31. Tam, D., MacLean, K.E., McGrenere, J., Kuchenbecker, K.J.: The design and field observation of a haptic notification system for timing awareness during oral presentations. In: Proceedings of the ACM SIGCHI Conference on Human Factors in Computing Systems (CHI '13), pp. 1689–1698. ACM, New York (2013). https://doi.org/10.1145/2470654.2466223
32. Zhao, S., Schneider, O., Klatzky, R., Lehman, J., Israr, A.: Feelcraft: Crafting tactile experiences for media using a feel effect library. In: Proceedings of the Adjunct Publication of the 27th Annual ACM Symposium on User Interface Software and Technology (UIST '14), pp. 51–52. ACM, New York (2014). https://doi.org/10.1145/2658779.2659109
33. Kittur, A., Chi, E.H., Suh, B.: Crowdsourcing user studies with mechanical turk. In: Proceedings of the SIGCHI Conference on Human Factors in Computing Systems (CHI '08), pp. 453–456. ACM, New York (2008). https://doi.org/10.1145/1357054.1357127
34. Heer, J., Bostock, M.: Crowdsourcing graphical perception: Using mechanical turk to assess visualization design. In: Proceedings of the SIGCHI Conference on Human Factors in Computing Systems (CHI '10), pp. 203–212. ACM, New York (2010). https://doi.org/10.1145/1753326.1753357
35. Chilton, L.B., Little, G., Edge, D., Weld, D.S., Landay, J.A.: Cascade: Crowdsourcing taxonomy creation. In: Proceedings of the SIGCHI Conference on Human Factors in Computing Systems (CHI '13), pp. 1999–2008. ACM, New York (2013). https://doi.org/10.1145/2470654.2466265
36. Mason, W., Suri, S.: Conducting behavioral research on amazon's mechanical turkBehav. Res. Methods**44**(1), 1–23 (2012). https://doi.org/10.3758/s13428-011-0124-6
37. Amazon.com Inc.: Amazon Mechanical Turk Requester Best Practices Guide (2015)
38. Downs, J.S., Holbrook, M.B., Sheng, S., Cranor, L.F.: Are your participants gaming the system? In: Proceedings of the ACM SIGCHI Conference on Human Factors in Computing Systems (CHI '10), p. 2399. ACM Press, New York (2010). https://doi.org/10.1145/1753326.1753688, http://dl.acm.org/citation.cfm?id=1753326.1753688
39. Modhrain, S.O., Oakley, I.: Touch TV: Adding Feeling to Broadcast Media (2001)
40. Kim, Y., Cha, J., Oakley, I., Ryu, J.: Exploring tactile movies: an initial tactile glove design and concept evaluation. IEEE Multimed. **PP**(99), 1 (2009). https://doi.org/10.1109/MMUL.2009.63, http://ieeexplore.ieee.org/lpdocs/epic03/wrapper.htm?arnumber=5255212

41. Abdur Rahman, M., Alkhaldi, A., Cha, J., El Saddik, A.: Adding haptic feature to YouTube. In: Proceedings of the International Conference on Multimedia (MM '10), pp. 1643–1646. ACM Press, New York (2010). https://doi.org/10.1145/1873951.1874310, http://dl.acm.org/citation. cfm?id=1873951.1874310

42. Eid, M., Alamri, A., Saddik, A.: MPEG-7 description of haptic applications using HAML. In: Proceedings of IEEE International Workshop on Haptic Audio Visual Environments and their Applications (HAVE '06), pp. 134–139. IEEE (2006). https://doi.org/10.1109/HAVE.2006. 283780, http://ieeexplore.ieee.org/lpdocs/epic03/wrapper.htm?arnumber=4062526

43. Eid, M., Andrews, S., Alamri, A., Saddik, A.E.: HAMLAT: A HAML-based authoring tool for haptic application development. In: LNCS 5024 - Haptics: Perception, Devices and Scenarios, vol. 5024, pp. 857–866 (2008). https://doi.org/10.1007/978-3-540-69057-3, http://www. springerlink.com/index/10.1007/978-3-540-69057-3

44. Kaaresoja, T., Linjama, J.: Perception of short tactile pulses generated by a vibration motor in a mobile phone. In: First Joint Eurohaptics Conference and Symposium on Haptic Interfaces for Virtual Environment and Teleoperator Systems, pp. 471–472. IEEE (2005). https://doi.org/10.1109/WHC.2005.103, http://ieeexplore.ieee.org/lpdocs/ epic03/wrapper.htm?arnumber=1406972

45. Brown, L.M., Kaaresoja, T.: Feel who's talking: using tactons for mobile phone alerts. In: CHI '06 Extended Abstracts on Human Factors in Computing Systems (CHI EA '06), p. 604. ACM Press, New York (2006). https://doi.org/10.1145/1125451.1125577, http://dl.acm.org/citation. cfm?id=1125451.1125577

46. Hoggan, E., Stewart, C., Haverinen, L., Jacucci, G., Lantz, V.: Pressages: augmenting phone calls with non-verbal messages, 555–562 (2012). https://doi.org/10.1145/2380116.2380185, http://dl.acm.org/citation.cfm?id=2380116.2380185

47. Israr, A., Zhao, S., Schneider, O.: Exploring embedded haptics for social networking and interactions. In: CHI '15 Extended Abstracts on Human Factors in Computing Systems (CHI EA '15), pp. 1899–1904. ACM Press, New York (2015). https://doi.org/10.1145/2702613. 2732814, http://dl.acm.org/citation.cfm?id=2702613.2732814

48. Felt, A.P., Egelman, S., Wagner, D.: I've got 99 problems, but vibration ain't one: A survey of smartphone User's concerns. In: Proceedings of the 2nd ACM Workshop on Security and Privacy in Smartphones and Mobile Devices (SPSM '12), p. 33. ACM Press, New York (2012). https://doi.org/10.1145/2381934.2381943, http://dl.acm.org/citation.cfm? id=2381934.2381943

49. Schuirmann, D.: On hypothesis testing to determine if the mean of a normal distribution is contained in a known interval. Biometrics **37**(3), 617 (1981)

Chapter 7
Tuning Vibrations with Emotion Controls

Abstract When refining or personalizing a design, we count on being able to modify or move an element by changing its parameters rather than creating it anew in a different form or location—a standard utility in graphic and auditory authoring tools. Similarly, in our study of haptic personalization mechanisms in Chap. 3, users preferred the *tuning* mechanism the most. For tactile vibration display, however, we lack knowledge of the human perceptual mappings which must underlie such tools. Based on evidence that affective dimensions are a natural way to tune vibrations for practical purposes, we attempted to manipulate perception along three emotion dimensions (*agitation*, *liveliness*, and *strangeness*) using engineering parameters of hypothesized relevance. Results from two user studies show that an automatable algorithm can increase a vibration's perceived *agitation* and *liveliness* to different degrees via signal energy, while increasing its discontinuity or randomness makes it more *strange*. These continuous mappings apply across diverse base vibrations; the extent of achievable emotion change varies. These results illustrate the potential for developing vibrotactile emotion controls as efficient tuning for designers and end-users.

7.1 Introduction

From cell phones to sensate suits, haptic technology has recently proliferated; studies routinely predict high utility for vibrotactile notifications in everyday life [1–4]. Adoption, however, has been slow. Advances in hardware theoretically allow sensations beyond undifferentiated buzzes, but even professional designers struggle to express memorable, aesthetically pleasing percepts by twiddling available engineering parameters. It can take years to develop a good intuition, and this knowledge is then hard to articulate or transfer. Personal or shared libraries of examples are currently the best mechanism; new expressive effects are often the result of modifying existing repertoires [5]. This is potentially a slow process, with most time spent laboriously exploring alternatives—a barrier to creative design, and the antithesis of improvisation. Perceptual controls that allow quick, direct modifications to sensations will be highly valuable in this process.

© Springer Nature Switzerland AG 2019

H. Seifi, *Personalizing Haptics*, Springer Series on Touch
and Haptic Systems, https://doi.org/10.1007/978-3-030-11379-7_7

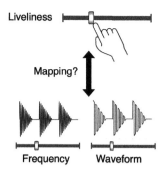

Fig. 7.1 Conceptual sketch of an emotion tuning control and its mapping to engineering attributes of vibrations

For end-users, personalization can improve utility and adoption of haptic signals [6, 7]. Consumers want to manipulate personal content more than ever [8–10]. The status quo is an immutable library, which provides users with a limited pre-designed set of effects to choose from. Given effective navigation, this helps; but given a choice, users have indicated preference for high-level controls that tune those predesigned effects to express a personal representation [11, 12].

In more mature domains, tools support varying levels of control and expertise. With Adobe Photoshop, one can manipulate pixel-level image features (crop, select a region, color fill), and overall perceptual attributes (brightness control, artistic filters) [13]. Adobe Lightroom provides photography enthusiasts with perceptual sliders to manipulate clarity, vibrance, saturation and highlights, which would otherwise require manipulating individual RGB pixel values in photo regions [14]. Instagram lets any smartphone user quickly choose perceptually-salient filters for more polished or customized images [15].

Manipulating vibrations brings similar needs. With existing tools, we modify *engineering* parameters: cropping part of a signal or changing its amplitude, waveform, or frequency at specific points along its timeline. With *sensory* controls, we could change perceptual attributes like roughness, speed, or discontinuity. Finally, *emotion* controls could address the mix of cognitive percepts that the vibration engenders. Here, an important question is what haptic controls would be most meaningful and useful to designers and end-users.

Past haptic studies suggest affective (emotion) dimensions to be an answer. While all three will be valuable for a professional designer, amateurs (whether a designer or an end-user) especially need the directness of emotion controls. Further, researchers have argued for the inherent neural link between touch and emotions, and the memorability of affective tactile signals [16–18]. Other findings point to the effectiveness of emotions as a framework for describing and accessing tactile sensations. In navigating a large vibration library, organized by a set of schemes including emotional as well as other descriptive perspectives (such as metaphoric or usage associations), users preferred and most often used the emotion scheme to find vibrations [12].

Together, these illustrate the importance of emotional traits as a target for meaningful vibrotactile design.

Throughout this chapter, we use the term *parameter* to signify vibration *attributes* that can be controlled. We aim to parametrize emotion attributes for manipulation and control (i.e., emotion *controls*), using the already manipulable engineering parameters (Fig. 7.1).

7.1.1 Research Questions, Approach and Contributions

In this chapter, we investigate the possibility of emotion controls for vibrations. We began from data indicating which emotion attributes users are most sensitive to: a previous analysis of user perception of a 120-item vibration library (*VibViz*) indicated primary alignments with *agitation*, *liveliness* and *strangeness* [19]. These became our candidate controls. For inclusion in design tools, such controls must further be automatable. This requires establishing a continuous mapping between the emotion attributes of interest and the manipulable engineering parameters of a display hardware (e.g., a C2 actuator [20]). The mapping must be consistent (or characterized) for a wide set of starting vibrations. Further, although not required for automatability, users can benefit from knowing the degree of emotion change, given a vibration's initial characteristics and the effect of adjusting one emotion control (e.g., *agitation*) on other emotion attributes (such as *liveliness* and *strangeness*).

We addressed the primary goal of automatable emotion controls through four subsidiary questions.

What vibrotactile engineering parameters influence primary emotion attributes?
Previous work showed the influence of engineering parameters on basic emotion dimensions of pleasantness and arousal. Here, we needed similar data for more nuanced emotion attributes. We selected a manageable set of starting-point "base" vibrations that represent the diversity in possible sensations; then determined influential engineering parameters from literature and experimentally, using sensory attributes (*e.g.*, *roughness*) as a middle step (Fig. 7.2). We derived a set of vibrations from the base examples by modifying those influential engineering parameters, and verified their impact in a study where participants rated *agitation*, *liveliness*, and *strangeness* of the vibration derivatives relative to the bases (Study 1). Towards this question, we contribute three sensory attributes that significantly impact perception of a vibration's three primary emotion attributes, as well as the engineering parameters that drive *them*.

Can we alter a primary emotion attribute of a vibration (e.g., its *liveliness*) **on a continuum by manipulating influential engineering parameters?**
We derived new stimuli from the base vibrations using three successively more extreme applications of the influential engineering parameters from the previous step. Then in Study 2, we examined whether these derivatives lay on a perceptual continuum between the emotion attributes and engineering parameters.

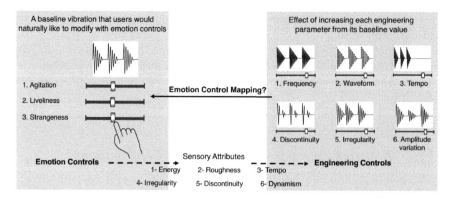

Fig. 7.2 Users mentally align vibrotactile sensations along several primary emotion attributes (left column). To exert direct control over these with design tools, we require a direct, automatable mapping from manipulable engineering parameters (solid line). To find this mapping, we used sensory attributes as a middle step—first establishing a link from emotion to sensory attributes, then from sensory to engineering parameters (dashed line)

How do characteristics of a base vibration impact a perceived change?
We examined how control effectiveness is amplified or minimized by properties present in a vibration starting point. We analyzed variations in the ratings provided in our two user studies for ten base vibrations that varied in their engineering characteristics, and showed that the mappings found for RQ2 hold for various vibration characteristics. We present qualitative descriptions of how these characteristics influence the extent of emotion change.

How independent are these emotion dimensions?
We analyzed correlation of ratings for the three dimensions, and tested for significant effects of engineering parameters on multiple emotion dimensions. We show that our proposed *emotion-engineering* mappings are not completely orthogonal: i.e., a change in an engineering attribute can impact perception of all three emotion dimensions to varying degrees.

In tackling these questions, we contribute:

- a process for parameterizing emotion attributes for control,
- two emotion controls and their mappings to engineering parameters, and
- directions for future research and development in haptic perception and design tools.

In the rest of this chapter, we first review related work (Sect. 7.2), then describe how perceptual controls can be used by designers and end-users (Sect. 7.3.1) and detail our process for identifying base vibrations and relevant vibrotactile engineering parameters Sects. 7.3.2 and 7.3.3. We detail the two user studies Sect. 7.4 and their results Sect. 7.5, discuss findings and three example tuning interfaces Sect. 7.6, then finish by outlining future avenues for research and tool design Sect. 7.6.4.

7.2 Related Work

7.2.1 Haptic Design, and Inspirations from Other Domains

Haptics designers commonly build on design guidelines or tool inspirations from more mature domains of design.

Design and personalization process: Built on existing theories of design thinking, MacLean et al. identified a set of major design activities and verified and characterized them for haptic experience design as follows: *browse*, *sketch*, *refine*, and *share* [21]. Design often starts by *browsing* existing collections to get inspiration, characterize the problem, and gather a starting set of examples. In *sketching*, designers quickly explore the design space by creating incomplete and rough sensations, making rapid changes to try alternative designs. Throughout the process, designers continuously *refine* a shrinking set of sensations to achieve a few final designs. Tweaking and precise aesthetic adjustments are the hallmarks of the refine activity. Finally, the sensations are *shared* with others to get feedback, reach target end-users, or disseminate design knowledge and contributions. In this framework, tuning controls facilitate the refinement process by expediting generation of salient alternatives for a given sensation.

Software and game personalization literature informs us about user motivations and desires. According to these, personalization increases enjoyment, self-expression, sense of control, performance, and time spent on the interface [9, 22, 23]. Ease-of-use and ease-of-comprehension in personalization tools engender take-up, while modifications are discouraged by difficulty of personalization processes [22, 24–27].

Building on these, we anticipate that an efficient *tuning* mechanism would enhance users' control and enjoyment of haptic notifications and improve their adoption rates among the crowds.

Intuitive authoring and personalization tools: Similarly, haptic authoring tools frequently incorporate successful paradigms from other domains. For example, Mango, an authoring tool for spatial vibrations like a haptic seatpad, is modelled after existing animation tools [28]. Exploiting music analogies, interfaces such as the Vibrotactile Score represent vibration patterns as musical notes [29]. Our inspiration for perceptual and emotion tuning controls comes from the visual and auditory domains. In music streaming platforms such as GooglePlay music, Musicovery, and MoodFuse, users can choose to search for songs based on key terms relating to mood or scenarios such as "keeping calm and mellow" or "boosting your energy" in addition to standard music genre categories [30–32]. Similarly, photo editing software such as Adobe Lightroom or Snapseed application utilize controls named to evoke emotion attributes such as "clarity" or "drama," which adjusts several pertinent features of the image (contrast, highlights) to create an effect [14, 33]. Among audio design tools, Propellerhead's "Figure" application provides audio presets such as "80's Bass" and

"Urban" as well as controls such as "weirdness" for creating and remixing music pieces [34].

These examples show the prevalence of perceptual controls for accessing and modifying stimuli in other modalities, and further highlight the gap in the haptic domain.

Stimuli design: Past research has drawn analogies between vibrotactile and audio signals to develop design guidelines and even hardware for haptics [20, 35–37].

Rhythm and pitch are important attributes of both audio and vibrotactile signals [35, 36]. Van Erp et al. designed 59 vibrations using short pieces of music while others developed crossmodal tactile and auditory icons based on common design rules [35–37]. In hardware design, voice coil actuators can take audio files as direct input and are commonly used in research for their high expressive range.

In this work, we benefit from these commonalities: we use an audio editing software, called Audacity, and a voice coil actuator (C2 tactor) to modify and display the vibration files [20, 38]. Further, we use the definition of tempo used in audio files and report its fit for users' perception of vibration's speed [39].

7.2.2 Affective Vibration Design

RQ1 builds on previous research in this area. Our own past work links the three proposed emotion dimensions to vibration sensory attributes; other studies provide guidelines linking sensory attributes to engineering parameters enabling the scheme laid out in Fig. 7.2.

VibViz library and five vibrotactile facets: In [12], we compiled five categories or facets of vibration attributes: (1) *physical* or engineering parameters of vibrations that can be objectively measured (e.g., duration, rhythm, frequency); (2) *sensory* properties (e.g., roughness); (3) *emotional* connotations (e.g., *exciting*); (4) *metaphors* that relate feel to familiar examples (e.g., *heartbeat*); and (5) *usage examples* or events where a vibration fits (e.g., *incoming message*). We designed a library of 120 vibrations for voice coil actuators (i.e., .wav files) and released a web-based interactive visualization interface (*a.k.a. VibViz*) that allows quick access to the vibrations through the five categories.

Here, we used the *VibViz* interface to choose a diverse set of basis vibrations from this library for our user studies.

Mapping engineering parameters to emotion and sensory attributes: In [19], we collected users' perception of the 120-item *VibViz* library according to the four perceptual facets of *sensory*, *emotion*, *metaphor*, and *usage example* attributes. We analyzed the ratings and tags provided, to identify the underlying semantic dimensions for these four facets. Results from factor analysis and correlation of tags, situated in different facets, linked sensory attributes of the vibrations to the other three facets. We summarized the results from that analysis into: (1) three emotion dimensions: *agitation, liveliness, strangeness*; and their correlation with (2) six sensory attributes: energy, roughness, tempo, discontinuity, irregularity, and dynamism (Sect. 10.1, Table 10.1 details these linkages).

Others linked vibration's engineering parameters to sensory attributes as well as to pleasantness and urgency [40–43]. Some general trends have emerged despite hardware dependence of specific engineering parameters and their reported threshold values: a vibration's energy depends on its frequency, amplitude, duration and waveform and sine waveform is perceived as smoother than a square wave [42, 44]. No definition exists for changing a vibration's tempo, (trivial) and discontinuity, irregularity, and dynamism. Also, past studies show that vibrations with higher energy, duration, roughness, and envelope frequency are less pleasant and more urgent [42, 43]. However, to our knowledge these studies do not go beyond pleasantness and urgency (*a.k.a.* arousal) to link more nuanced emotion attributes to engineering parameters.

In this chapter, our objective is to develop *emotion-engineering* mappings for our three emotion attributes, thereby creating a path through which we can control these cognitive dimensions—which up to this point we have been able to perceive and analyze with, but not produce at will [12, 19].

7.3 Starting Points: Use Cases, Initial Vibrations and Linkages

To address our research questions, we carried out three initial steps. First, we established a set of guiding use cases to frame our studies. Then, as a starting point for tuning, we chose a vibration subset from the *VibViz* library with relevant diversity. Finally, we estimated initial linkages of the emotion attributes to engineering parameters using past literature and our own pilot studies.

7.3.1 Design and Personalization Use Cases

In two exemplar use cases, emotion controls facilitate otherwise cumbersome design and personalization tasks.

Tuning a vibration set for a game (Fig. 7.3a): Alex, a haptics designer, is developing a set of vibration effects for different scenes and interactions in a new multimodal game. While talking to stakeholders, he refines some of the sensations to be more "alien", "fun", or "agitating", trying for a distinct yet coherent sensation experience. He iteratively adjusts emotion and engineering controls for several vibrations in the game set, testing each alternative quickly and comparing the feel with the rest of the vibrations in the set.

Personalizing daily notifications (Fig. 7.3b): Sarah often does "interval workouts"— alternating fast and slow running pace for pre-set durations. She needs clear notifi-

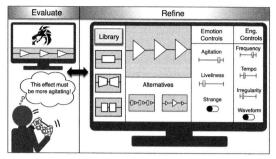

(a) Haptic **design** inevitably involves several rounds of evaluating sensations (left) and refining them (right). With emotion controls, designers could efficiently explore the affective design space around an example or starting point.

(b) **Personalization:** End-users untrained in haptics could efficiently personalize vibration notifications in situ, during or after use, by applying emotion filters to preset vibrations.

Fig. 7.3 **Use cases** for tuning vibrations' characteristics, using parameters aligned with users' cognition and design objectives: for both cases, controls based on emotion attributes enable "direct manipulation" from the user perspective

cations when an interval starts, or ends, both differentiable from other notifications even when she's strenuously exercising. Recently she has installed an application that lets her select the events triggering a notification on her smartwatch and associate them with vibrations from a list. She can preview and apply alternative feels for a vibration (e.g., a more *lively* version) by quickly tapping on available emotion filters.

We note that while both designers and end-users may wish to tweak a single or set of sensations, user *groups* may have different needs. We anticipate that when the latter customize sensations for their own use, they will prefer simple and quick adjustments with intuitive controls. Conversely, the former may need to achieve more polished or generalizable results, and will need finer-tuned control over emotion as well as engineering controls.

7.3.2 Choosing Basis Vibrations

To develop an emotion control that can tune any given vibration, one needs to either study a large set of vibrations with many attributes, or examine a smaller set in a systematic way. The first approach requires extensive data collection and large-scale (e.g., crowdsourced) experimental methods that are currently difficult with haptics [45]. We chose the second approach, using rhythm to structure our investigation as past research report it to be the most salient perceptual parameter for determining vibration similarity [46].

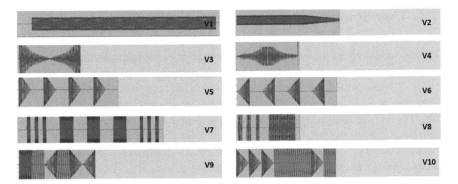

Fig. 7.4 Ten basis vibrations (five pairs) from the *VibViz* library, selected for our studies as tuning starting points. Each row represents a vibration pair that shares unique rhythm and envelope attributes not found in other pairs. As an example, V9 and V10 both have several connected pulses with various envelopes (constant, rampup, or rampdown)

Two authors independently chose a representative subset of *VibViz* vibrations which varied in rhythmic features, and consolidated them into a 17-item set. We further narrowed these to five vibration pairs, with each pair representing a rhythm family (Fig. 7.4), to examine consistency of the *tuning* results within and between the paired members.

7.3.3 Identifying Influential Engineering Parameters

In a two-step process, we first identified an emotion to sensory (*emotion-sensory*), then a *sensory-engineering* mapping.

Emotion-sensory mapping: In a previous work, we identified sensory attributes correlated with each emotion attribute [19]; see Sect. 10.1, Table 10.1 for a summary. Based on those results, we selected six attributes for further investigation: energy, roughness, tempo, discontinuity, irregularity, and dynamism.

Sensory-engineering mapping: We derived relevant engineering parameters for energy and roughness from the literature but did not find prior work defining tempo, discontinuity, irregularity, and dynamism. For these, we manually and iteratively altered our initial 17 vibration *.wav* files using the Audacity audio editing tool [38], testing candidates in small pilots. We tested various applications of these sensory attributes until we converged at six potentially influential engineering parameters (frequency, waveform, tempo, discontinuity, irregularity, and amplitude variation) for further investigation in user evaluations (see Sect. 10.2, Table 10.2 for more details on our sensory to engineering mappings).

7.4   User Studies

Having identified a set of potentially influential engineering parameters, we sought
continuous mappings from them to emotion attributes for a given base vibration
(RQ1-4). Here, we followed an exploratory research process involving a pilot and two
user studies, with each study informing the parameters and study design for the next
one. In all of the studies, participants rated stimuli derived from a base vibration, on
agitation, *liveliness*, and *strangeness*. In Study 1, we verified that a mapping existed
between the emotion and engineering parameters noted in Sect. 7.3.3 and in Study 2
we tackled the mappings' continuous nature.

7.4.1 Overview of the User Studies

Pilot Study: We established our study protocol in a pilot study with 10 participants, where we studied six derivatives for each base, designed by modifying one
of the six engineering parameters identified in Sect. 7.3.3 (frequency, waveform,
tempo, discontinuity, irregularity, and amplitude variation). Results indicated two
top-performing engineering parameters for each dimension: for *agitation*: waveform and frequency; *liveliness*: waveform and tempo; *strangeness*: discontinuity and
irregularity.
Study 1—Verifying influence of engineering parameters on emotion attributes: We
examined the effect of the top-performing engineering parameters on the emotion
attribute in a formal user study and explored whether one can achieve a more pronounced emotional effect by applying changes to both top-performing parameters.
Study 2—Evaluating continuity of engineering-emotion mapping: The next step
was to establish continuity in a mapping from engineering parameters to emotion
attributes (RQ2), by examining the impact of successively more extreme applications of the engineering parameter combinations that were found to be influential in
Study 1, namely: frequency+waveform, and irregularity+discontinuity. We investigated the effect of an increase in frequency+waveform and irregularity+discontinuity
on *agitation*, *liveliness*, and *strangeness*.

7.4.2 Methods

Studies 1 and 2 (as well as the pilot) shared apparatus and procedure. The studies
differed in stimuli set and size. Both studies were approved by the ethics review
board at the University of British Columbia.
Stimuli

Study 1: We utilized all 10 base vibrations (5 pairs), creating eight derivatives for each
as follows: (a) the base vibration itself, as a statistical control; (b) six derivatives per

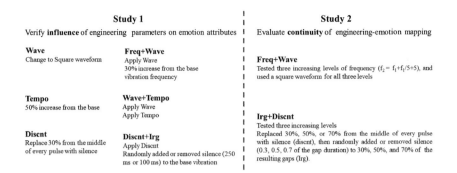

Study 1		Study 2

Study 1
Verify **influence** of engineering parameters on emotion attributes

Wave
Change to Square waveform

Freq+Wave
Apply Wave
30% increase from the base
vibration frequency

Tempo
50% increase from the base

Wave+Tempo
Apply Wave
Apply Tempo

Discnt
Replace 30% from the middle
of every pulse with silence

Discnt+Irg
Apply Discnt
Randomly added or removed silence (250
ms or 100 ms) to the base vibration

Study 2
Evaluate **continuity** of engineering-emotion mapping

Freq+Wave
Tested three increasing levels of frequency ($f_2 = f_1 + f_1/5 + 5$), and
used a square waveform for all three levels

Irg+Discnt
Tested three increasing levels
Replaced 30%, 50%, or 70% from the middle of every pulse
with silence (discnt), then randomly added or removed silence
(0.3, 0.5, 0.7 of the gap duration) to 30%, 50%, and 70% of the
resulting gaps (Irg).

Fig. 7.5 Overview of the engineering parameters and evolution of their functional implementation to achieve control over the three emotion attributes in Studies 1 and 2. Values listed under each engineering parameter indicate changes applied to the base vibrations in the associated study. "Freq", "wave", "discnt", and "irg" denote frequency, waveform, discontinuity, and irregularity respectively

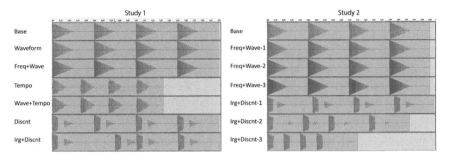

Fig. 7.6 An example of vibration derivatives in Study 1 and 2 (designed for base vibration V5). Increasing frequency is represented through increased image color saturation. Increasing tempo (i.e., rhythmic rate) resulted in shorter signals as a side effect. Discontinuity and irregularity+discontinuity are implemented by adding silent periods (represented as zero amplitude), and by varying the duration of these silent periods

base, representing change in waveform, tempo, discontinuity, frequency+waveform, waveform+tempo, and irregularity+discontinuity (see Figs. 7.5, 7.6), and (c) a randomly chosen duplicate of one of these seven, to assess rating reliability.

This resulted in a total of 90 vibrations (10 base and 80 derivatives) rated in comparison to the base vibrations by each participant—i.e., 80 comparisons.

Study 2: We included eight derivatives for each of the 10 base vibrations: (a) the base vibration itself, (b) three levels of frequency+waveform, (c) three levels of irregularity+discontinuity, and (d) a randomly chosen duplicate of one of these seven.

As for Study 1, this resulted in 90 vibrations (10 base, 80 derivatives) rated by each participant—80 comparisons.

For the frequency+waveform derivatives, the frequency increase at each level was based on the Weber's JND law ($f_2 = f_1 + \dfrac{f_1}{5} + 5$). Waveform did not change across the

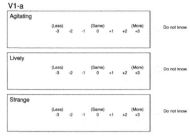

(a) Apparatus for user studies

(b) Rating interface showing one vibration derivative and three Likert item ratings representing the three emotion attributes.

Fig. 7.7 Experimental setup for the pilot and Studies 1 and 2. The rating interface shown in **b** appears on the computer screen in **a**

three levels. For the irregularity+discontinuity derivatives, we first applied discontinuity by removing 30, 50, and 70% from the middle of each pulse in the vibration. To systematically vary irregularity, we then randomly added or removed silence from the first 30, 50, or 70% of the resulting gaps, with the amount of silence proportional to the duration of the gap (30, 50, and 70% of the gap duration respectively, which translated to values between 0 and 0.4 ms—Fig. 7.5).

Participants: We recruited 20 (12 females, 18 native English speakers), and 22 (15 female, 19 English) participants for Study 1 and 2 respectively by advertising on a North American university campus ($15 compensation for a 1-hour session). Individuals could only attend one of the two user studies. Participants reported no exposure to haptic signals other than vibration buzzes on their cellphones. Study 1 and 2 lasted an average of 35 and 45 mins respectively.

Apparatus: To display the vibrations, we used a C2 tactor, connected to an amplifier and a laptop. Each base vibration and its derivatives were placed in a separate desktop folder visible on the laptop screen. The rating interface was an online questionnaire with each page representing all the derivatives and required ratings for one of the base vibrations (randomized order). Each question on a page displayed the name of the derivative (e.g., V1-a) and three Likert item ratings (-3 to $+3$) for *agitation*, *liveliness*, and *strangeness* (Fig. 7.7b). A rating of -3 indicated that a derivative had considerably less of an emotion attribute compared to the base (i.e., less agitating or negative influence of the *engineering* parameter), while $+3$ indicated having more of the emotion attribute (i.e., more agitating or positive influence). Participants played the vibration files on the laptop, provided their ratings on the survey, and listened to pink noise through headphones to mask any sound from the actuator.

Procedure: Study sessions were held in a private, closed-door room and started with a short interview. After asking for the participant's demographics, the experimenter asked them to imagine and define an *agitating*, *lively*, or *strange* vibration using their own free-form words and typed their responses on a computer. To calibrate on common definitions, the experimenter then provided a verbal definition of the three emotion terms with short lists of adjectives drawn from emotion synonyms in [19] and asked them to use these synonyms in the remainder of the study:

lively: happy, energetic, interesting
agitating: annoying, urgent, angry, uncomfortable
strange: odd, unfamiliar, unexpected

The experimenter described these synonyms again whenever the participants asked for them but in most cases, these synonyms were well-aligned with the participants' initial descriptions for the emotion attributes (see Sect. 10.3, Table 10.3). The rating task consisted of feeling all the derivatives for a base vibration first, then providing three ratings for each derivative to indicate whether it was more/less *agitating*, *lively*, and *strange* than the base vibration or to mark a rating with "do not know". Participants held the tactor between the tip of the fingers (Fig. 7.7), and rated each derivative once (randomly ordered) while having access to its base vibration as well as all other derivatives in that set (placed in a folder on the experimenter's laptop) throughout the experiment. In a post-interview, the experimenter asked for and recorded participants' definition of the three emotion terms, to identify any changes in the emotion definitions as a result of feeling the vibrations. At the end, the experimenter addressed any questions about the study such as its potential use cases and the engineering parameters used in the experiment.

7.4.3   Analysis

Replaced Values: Out of over 10,000 ratings collected, we received a small number (five in Study 1, six in Study 2) of "do not know" responses. We replaced these with the median of the other ratings for the corresponding derivative.

Duplicate Trials: The median of rating differences between a derivative and its duplicate (inserted to estimate reliability—see Sect. 7.4.2) was 0 and 0.5 (7-point scale) in Study 1 and 2 respectively. We therefore removed ratings for the duplicate derivatives for the rest of our analysis.

Nonparametric Factorial Analysis (ART): We then performed a full factorial analysis to test for the effects of the engineering parameters. Because this involved multiple nonparametric factors, we used the Aligned Rank Transform (ART) for nonparametric factorial analyses [47]. ART was designed for and has been used by many as a multifactor nonparametric alternative for ANOVA. It applies a rank transformation on the rating data [48], then runs an ANOVA test on the ranks. Thus, results from ART are interpreted similarly to the ANOVA results. For each study, we ran the test on the ratings for *agitation*, *liveliness*, and *strangeness* separately, using two factors of engineering parameter (7 levels) and base vibration (10 levels). Since ART is an omnibus test, we used Tukey's posthoc analysis with corrected p-values for multiple comparisons with an alpha level of 0.05.

7.5   Results

We first present qualitative descriptions collected in the pre and post interviews, then show minimally processed rating data, and present our ART analysis with respect to our research questions (RQ1-4).

7.5.1   Verbal Descriptions for Emotion Attributes

We aggregated the emotion descriptions collected from the participants in the semi-structured pre- and post-session interviews for Studies 1 and 2 as follows. We extracted adjectives (e.g., irritating) and noun phrases (e.g., short pulses), consolidated synonyms (e.g., fast and agile), and counted total usage instances for each adjective across the participants. For example, we coded the Study 2 definition of a *lively* vibration by P18 ("more intense and faster vibrations") as strong (+1) and fast(+1); then summed with similar adjectives from other participants (See Sect. 10.3, Table 10.3).

For all three emotion attributes, in the pre-interview participants mostly used descriptive emotion words when we asked them to define these concepts as they might be expressed as vibrations, in their own words. Given the same question in the post-interview, they generally drew upon sensory definitions such as vibration structure and feel.

The pre-interview produced several patterns. Both *agitating* and *strange* vibrations (considered in the abstract) were labelled with adjectives typically considered unpleasant and negative (e.g., unexpected and unfamiliar for *strange* vibrations, irritating and nervous for *agitating* ones). In contrast, *liveliness* was associated with positive attributes such as energetic, happy and pleasant.

In the post-interview, sensory definitions for *agitation* overlapped in content with both *liveliness* and *strangeness*, but the latter two did not share any descriptions (per participant or when aggregated). Specifically, *agitating* or *lively* vibrations were both described to be strong and fast, but they differed in other ways: *liveliness* was linked to short pulses and a rhythmic pattern while long, non-rhythmic, and irregular vibrations were considered *agitating*. *Strange* vibrations shared part of the *agitation* space, being likewise described as irregular and off-rhythm.

7.5.2   Ratings

We collected a total of 10,080 emotion attribute ratings for Study 1 and 2 vibration derivatives. Figure 7.8 shows these as boxplots for *agitation*, *liveliness*, and *strangeness*.

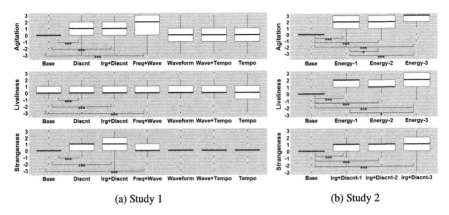

(a) Study 1   (b) Study 2

Fig. 7.8 Boxplot of *agitation*, *liveliness*, and *strangeness* ratings for the base vibration and vibrotactile derivatives representing changes in the engineering parameters in Study 1 and 2. Starred lines mark significantly different pairs of conditions, with *** and * indicating significant results at $p < 0.0001$ and $p < 0.05$ respectively. "Freq+Wave", "Wave+Tempo", "Discnt", and "Irg+Discnt" denote frequency+waveform, waveform+tempo, discontinuity, and irregularity+discontinuity respectively

To denote patterns of the ratings pertaining to all 10 base vibrations and 7 engineering parameters, we then visualized average ratings for each vibration derivative (Fig. 7.9). Average ratings of -3, 0, and $+3$ indicate negative, zero and positive influence of an emotion attribute on the derivative compared to the base vibration.

7.5.3   RQ1: Impact of Engineering Parameters on Emotion Attributes (Study 1)

The first research question's objective was to establish which engineering parameters (which we are able to manipulate) can influence perception of emotion attributes. ART analysis (Sect. 7.4.3) of our Study 1 data showed a significant main effect of engineering parameter and a main effect of base vibration on the ratings for all three emotion attributes. A posthoc Tukey's test determined which engineering parameters were significantly different from the base.

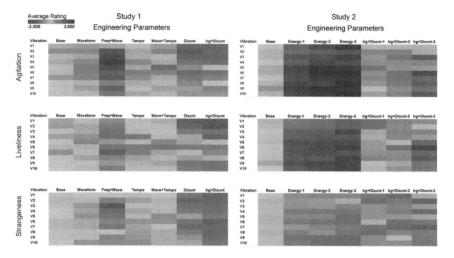

Fig. 7.9 Average ratings of the emotion attributes in response to variation of engineering parameter combinations (subfigure columns) for the 10 base vibrations (subfigure rows) in Studies 1 and 2. Influence of the engineering parameters on the base vibrations for that emotion attribute is denoted by color: blue is negative (bounded by average rating of −3.0, intense blue), gray is neutral (0), and red shows a positive influence (bounded at +3.0). Column saturation thus indicates strong *influence* (positive or negative) of an engineering parameter, whereas row saturation indicates *susceptibility* of that vibration to being influenced. Consistent color and saturation in a column indicates a consistent perception regardless of the base vibration; color variation suggests dependency on the base vibration

Figures 7.8, 7.9 illustrate the outcomes. Specifically, applying frequency+waveform, discontinuity, or irregularity+discontinuity to the base vibrations resulted in a statistically significant increase in all three emotion dimensions while applying waveform, tempo, or waveform+tempo had minimal impact on the perceived emotions.

In Fig. 7.9, the columns representing no statistical significance (waveform, tempo, and waveform+tempo) show either low emotion change (grey or low saturation cells) or inconsistent change for different base vibrations (color variations). In contrast, the majority of cells for the statistically significant parameters (the frequency+waveform, discontinuity, and irregularity+discontinuity columns) show high emotion influence (highly saturated cells). Further, frequency+waveform resulted in the most consistent perception for *agitation* and *liveliness* while irregularity+discontinuity led to consistent results for *strangeness*.

In summary, Study 1 succeeded in highlighting possible control paths towards all three emotion attributes. Notably, the *agitation* and *liveliness* attributes shared the same engineering parameters in these results. We further investigate an overlap in their continuous mappings in Study 2.

7.5.4 RQ2: Evidence of Continuity of the Engineering-Emotion Mappings (Study 2)

In Study 2, we investigated mapping continuity, using three successively more extreme applications of the influential engineering parameter combinations to create the vibration derivatives.

Frequency+waveform

In analyzing the stimuli for Study 2, we noted that the three increasing levels of frequency+waveform resulted in very different actuator output energy (actuator acceleration measured as m/s^2) depending on the actuator's frequency response curve (peak at $f = 275\,Hz$) and a base vibration's frequency. Thus, in our analysis, we ordered the vibration derivatives used in Study 2 according to the frequency of the tactor's peak response to dictate three increasing energy values. That is, the energy sequence of $[base, energy_1, energy_2, energy_3]$ simply swapped the order of the 2nd and 3rd frequency+waveform derivatives: $[f_0 = 200\,Hz, f_1 = 245Hz, f_3 = 352\,Hz, f_2 = 289\,Hz]$.

Running ART and Tukey's posthoc on the energy-ordered data showed a significant effect of energy on all three emotion attributes. Specifically, for *agitation* ratings, this resulted in significant differences between all three energy levels (borderline significance for the revised $energy_1$ and $energy_2$, $p = 0.1$).

For *liveliness*, Tukey's posthoc showed a borderline difference between $energy_1$ and $energy_3$, $p = 0.08$. For *strangeness*, the test resulted in significant differences between the derivatives and the base, with no difference between three successive energy derivatives.

Figure 7.9's visualization is consistent with these results. *Agitation* and *liveliness* cells show an increase in emotion change (higher saturation) for higher energy levels. We note, however, that *liveliness* cells are less saturated than the *agitation* ones for the same derivative, suggesting that these energy changes impacted *liveliness* less than *agitation*. *Strangeness* cells show an increase in saturation for three all energy levels but the effect is smaller and less consistent than the ones for *agitation* and *liveliness*.

Irregularity+discontinuity

ART showed significant main effects of the irregularity+discontinuity for all three emotion attributes. For *agitation* and *liveliness* ratings, Tukey's posthoc resulted in significantly lower ratings for all three successive energy levels (borderline significance for the $Irg + discnt_2$ and $Irg + discnt_3$ in *liveliness* ratings, $p = 0.1$). For the *strangeness* ratings, the three levels of irregularity+discontinuity were significantly different from the base but not different from each other.

7.5.5   RQ3: Impact of Base Vibrations on Emotion Attribute Ratings

Our ART analysis suggested that the base vibrations varied in their emotion change after applying the engineering parameters—a significant main effect of base vibration in both studies. Figure 7.9 depicts differences in the emotion ratings for the 10 base vibrations.

Agitation and liveliness: For all the base vibrations in both studies, applying some level of frequency+waveform (or *energy*) tended to increase their perceived *agitation* and *liveliness* (grey to red colors). However, the extent of increase varied for different base vibrations. These differences are more pronounced in Study 1 results but are resolved after the energy re-ordering in Study 2.

   To see if a vibration's base rhythm contributed to the ratings, we examined consistency of the ratings for the paired vibrations (Sect. 7.3.3) but only found one notable instance. V1 and V2, paired for being continuous and flat, received lower *liveliness* ratings that the other vibrations even with the Study 2 energy reordering.

Strangeness: All the base vibrations became more *strange* after applying irregularity+discontinuity and discontinuity. However, in some cases, the boost was minimal (low saturation cells). In Study 2, some base vibrations showed a consistent albeit gradual increase in *strangeness* (e.g., V4 and V7), but the majority did not. This is consistent with the statistical results (significant main effect but no pairwise significance) in Sect. 7.5.4. Examining the paired vibrations did not yield any apparent link between the *strangeness* ratings and rhythm patterns of the base vibrations.

7.5.6   RQ4: Orthogonality of Emotion Dimensions

Correlation among the three emotion dimensions: A Spearman's rank correlation test was positive for *agitation* and *liveliness* ratings (strong for Study 2 ($r = 0.67$), and weak in Study 1 ($r = 0.39$)). For Study 1, Spearman's test also revealed a moderate correlation between *agitation* and *strangeness* ($r = 0.44$).

ART results: According to the RQ1-2 results, frequency+waveform and irregularity+discontinuity impacted all three emotion attributes, however their effect was more prominent for some emotion attributes than for others. While *agitation* and *liveliness* consistently increased with energy in both studies, the effect of irregularity+discontinuity was not consistent: these parameters tended to significantly increase *agitation* and *liveliness* ratings in Study 1 but significantly decreased them in Study 2. In both studies, *strangeness* ratings were increased by applying irregularity+discontinuity and frequency+waveform, with the impact of irregularity+discontinuity being stronger.

7.6 Discussion

After a summary of our findings, we discuss automatability of the emotion controls given these results and reflect on our study approach, and finally present three example interfaces supporting our design and personalization use cases.

7.6.1 Findings

Evidence of mapping from engineering parameters to emotion attributes (RQ1 and RQ2): We found a set of engineering parameters that can increase perception of *agitation*, *liveliness*, and *strangeness* for a given vibration. Specifically, our results suggest a linear relationship between *agitation*, *liveliness* and the actuator's output energy. Adding irregularity and discontinuity to a vibration increases its *strangeness* but the effect does not increase with the degree of discontinuity and irregularity.

Differences observed for the base vibrations (RQ3): The extent of emotion boost depends on the characteristics of the base vibration. We found that differences in *agitation* and *liveliness* boosting were best described by the actuator's output energy, as evinced by the improved monotonicity of relationship in Study 2 versus Study 1. Rhythm and envelope played a secondary role for *liveliness*, where continuous and flat base vibrations (V1, V2) received a lower boost than did the other bases for a similar increase in energy. V7, with a symmetric rhythm of short and long pulses, was among the most lively vibrations for different energy levels.

Strangeness ratings were mixed. This may have been due to using random values in our irregularity derivatives: sometimes this produced a regular rhythm (e.g., irg+discnt-3 for V1), and elsewhere, noticeably irregular beats.

Orthogonality of the emotion controls (RQ4): Our results suggest two relatively independent emotion controls for vibrations: one that modifies *agitation* and *liveliness* jointly, and another one for *strangeness*. Our results also suggest a subtle distinction between *agitation* and *liveliness* (e.g., impact of base rhythm and qualitative descriptions), which need further examination. Finally, we saw that a change in one emotion attribute can influence perception of others.

We next discuss potential automatability of emotion controls, given these results.

7.6.2 Automatable Emotion Controls and Study Approach

Our studies show that at least one automatable solution exists. Since our goal was to verify the existence of at least one mapping, we chose the most promising sensory to engineering mappings based on the literature and our pilots (Sect. 3.3) instead of doing an exhaustive search of the sensory and engineering parameter spaces in formal user studies. The results confirmed the viability of our proposed mapping for a diverse set of base vibrations. The mapping, however, is neither orthogonal nor uniform. The extent of change along the emotion dimension can vary for different vibrations, and moving a vibration along one emotion dimension can impact its other emotion attributes. These qualities are not surprising; they exist in other domains and do not undermine the effectiveness of the controls. As an example, in Adobe Lightroom, increasing the "shadows" does not change every photo to the same degree. Further, the effect of adjusting "shadows" on a photo's "vibrance" is not always predictable.

We used a top-down approach in designing the emotion controls. We started with a set of emotion attributes, then devised a mapping to the engineering parameters. A bottom-up approach would have required developing a set of engineering controls, then building higher level controls based on emerging usage trends over time. This would have necessitated long-term usage or access to crowds to aggregate usage patterns, and the resulting controls may require background knowledge (e.g., "highlights" vs. "whites" sliders in Adobe Lightroom). Given a lack of access to the crowds [45] or a large established haptic design community, we chose the top-down approach to find an existence proof rather than an optimized solution. This process is not the only possible way, nor are these mappings necessarily unique. Over time, we anticipate that triangulation of different approaches will lead to the best results.

Exposure to vibrations led to more concrete descriptions for the emotion terms which in turn highlighted next steps for the engineering-emotion mappings. We relied on qualitative descriptions to gain perspective on practical significance of our quantitative results. Interestingly, participants became more articulate after a relatively short exposure to the vibrations. At the start, they described the three emotion dimensions mostly with other emotion words but in the post-interview, they commonly referred to the sensory and engineering parameters of the vibrations (e.g., strong, frequent pulses). In most cases, the participants' definitions were consistent with their ratings and our hypothesized engineering parameters. In one exception, *lively* vibrations were commonly described as "fast" but tempo and waveform+tempo were not effective. We increased tempo based on the definition available for audio tracks because of the perceptual and design commonalities between the vibrotactile and auditory modalities. But our results suggest that this definition is not aligned with users' perception and calls for a more effective model for the perception of speed in vibrations.

7.6.3   What Do These Results Enable? Revisiting Our Use Cases

Our motivation for this research was to empower the creation of haptic design and personalization tools. Here, we discuss three interface concepts, informed by our results, as a vehicle to relate our findings back to the design and personalization scenarios in Sect. 7.3.1 and reflect on their underlying parameters.

Fig. 7.10 Instagram for vibrations. Users would sketch a new vibration through a simple interface (e.g., by tapping on the phone screen, recording their voice, drawing a rhythm, etc. (left)) and then stylize it by applying emotion filters (right)

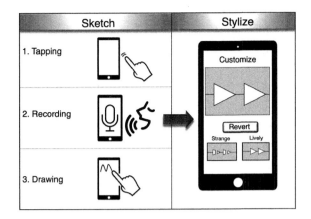

Vibration Instagram: Our personalization use case in Sect. 7.3.1 calls for a simple interface where ordinary users can apply a set of predefined effects to a given vibration (chosen from default vibrations on their phone or sketched with a simple tool (Fig. 7.10). All the effects use the same binary control, and must be perceptually salient but may not rely on pre-defined emotional connotations. Users have no access to underlying engineering parameters.

Grounded in our results, one can create at least two emotionally meaningful filters (*agitating/lively*, and *strange* effects), and several other filters representing alternative applications of an engineering parameter (e.g., different levels of irregularity+discontinuity for distinct *strange* sensations).

Emotion Toolbox: In our design use case (Sect. 7.3.1), Alex could benefit from an add-on toolbox to his vibrotactile authoring tool(s) that provides full access to the available emotion controls (e.g., switch, or a slider denoting the binary or continuous nature of the possible emotion change) and their underlying engineering parameters. Ideally, the interface will allow designers to define new proprietary controls and map them to the engineering parameters.

Grounded in our results, we can now create a toolbox with slider controls for *agitation*, *liveliness*, and a switch control for *strangeness* (Fig. 7.11). In a "details" layer, designers can see default engineering settings for each control, change the preset values for the controls, and access other engineering controls not used in current attribute definitions.

Vibrotactile Palette Generator: In designing vibration derivatives for our studies, we noticed several cases where seeing a *palette* of vibrations, each of which are perceptually distinct from the others but share a common theme, would be useful for personalization and design. For example, an end-user may wish to create a homogeneous set of wakeup alarms with increasing *agitation* for each snooze round. These cases benefit from an interface that can automatically generate a set of derivatives based on an input vibration along a predetermined emotion dimension (Fig. 7.12). Users would have access to one type of multi-level discrete control for all the emotion

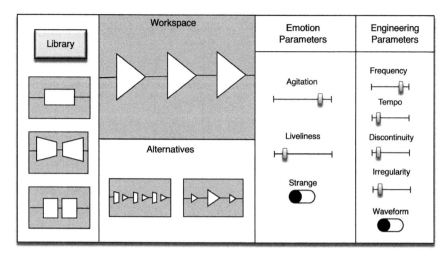

Fig. 7.11 Emotion toolbox. Designers would be able to start from a vibration in their library (left panel), use high-level emotion controls (third panel), and override default engineering presets as needed (right panel). Promising candidate would be saved to the bottom alternatives panel

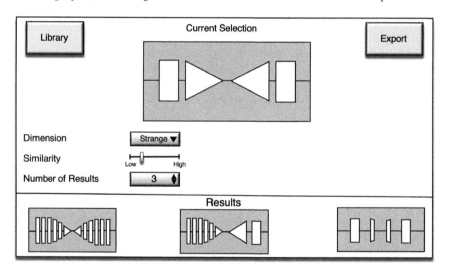

Fig. 7.12 Vibrotactile palette generator. Users would select a base vibration (from a library), determine the emotion dimension for derivatives, their similarity, and number of derivatives. The system would automatically create the derivatives on demand based on a predefined algorithm(s)

dimensions, denoted by meaningful emotion labels to guide derivative generation, and subsequent control over semantic parameters of the underlying algorithm (e.g., number and similarity of the derivatives).

This interface concept would exploit our findings, in the form of a palette generator along the *agitation/liveliness* dimension and another along the *strangeness*) dimension and vary energy and irregularity+discontinuity to derive perceptually distinct sensations with predictable emotion impact. Although applicable to both dimensions, this interface is mostly appropriate for emotion attributes such as *strangeness* which has known engineering parameter(s) but no linear effect of the parameter.

Reflections on supporting tuning scenarios: Our use cases in Sect. 7.3.1 vary in two parameters: (1) target users—designers versus naive users, and (2) tuning task—tweaking a single versus a set of sensations. We can now use these interface concepts as a basis for discussing how tuning scenarios, varying on these two parameters, may be supported through interface design choices.

Target Users: Ease-of-use and design control are typical interface properties of interest to both of the demographics we have identified—designers and end-users [11]. Our results indicate feasibility of achieving balances for these two properties and suggest ways they might be embodied. Specifically, Vibration Instagram limits vibration alternatives and control (fixed binary presets) to achieve simplicity and efficiency of use, and thus lends itself to users who may have less design knowledge or willingness to tackle a learning curve. In contrast, Emotion Toolkit emphasizes flexibility and control for designers who may find its power and subtlety a worthwhile tradeoff for complexity, as with users of other expert tools. Finally, Vibrotactile Palette Generator provides a middle ground through semantic control over the underlying algorithm, making it suitable for both designers or tech-savvy end-users.

Vibration Tuning Task: In modifying sensations, users may wish to tweak a single item or a set of them. Our proposed interfaces support one or both of these workflows—another factor in what might suit them for a particular demographic or context. Vibration Instagram allows tuning just one sensation at a time, while Vibrotactile Palette Generator's purpose is to facilitate creating sets of related sensations. Emotion Toolkit supports both tasks; the designer can tweak a single vibration with the sliders but can also access and/or save to the vibration set using the library and alternatives panels.

7.6.4 Future Work

There are several avenues for extending our work. First, our results mostly reflect North American perception of emotions; Verifying and extending these mappings for other cultures is an open area for future research. Second, to manage scope, we focused on the mappings that increase perception of the three emotion attributes, leaving investigation of the effect in the opposite direction (decreasing emotion percept) to future studies. Third, modelling the mappings we identified between engineering parameters and the three emotion dimensions using regression and other statistical techniques, would be a critical step for automating the operation of emotion controls.

Fourth, we did not study utility of these controls for designers. Thus, to complement tool development efforts, future studies can examine these controls in situ and define relevant evaluation metrics.

We close by pointing to other conceptual approaches for moving a vibration in the emotion space which can complement and/or extend our work. Conceptually, our controls enable *extrapolation*; they start from one existing sensation and generate a new one based on a set of rules. Other frameworks should also be considered. For example,

The system recommends alternative (but existing) vibrations from a library that have the desired emotion attribute(s) (e.g., are more *lively*) but otherwise share structure and engineering parameters with the base vibration. The user "navigates" to the alternative vibrations in the library, guided by the system recommendations.

The system creates a new vibration in between a starting base vibration and another with the desired emotion attributes. To make a vibration more *lively*, for example, it would interpolate between the base and a *lively* vibration. The interpolation ends could be specified by the user or automatically selected from a library. These approaches pose different challenges and opportunities. Once applied in a design tool, they can complement one other to provide a rich toolset for designers, or a seamless personalization mechanism for end-users.

7.7 Conclusion

Inspired by existing authoring tools in visual and auditory domains, this chapter examined design of emotion and perceptual controls for haptics and took a critical step towards parameterizing the emotion attributes for control and design. In particular, we presented our process for tackling many challenges that are specific to automatable emotion controls, namely: establishing continuous mappings along specific emotion attributes, identifying consistent effects across various base vibrations, and finding orthogonal or consistent effects across the emotion attributes.

Through user study results we showed that emotion controls can be created automatically, and proposed a mapping from the emotion attributes of *agitation, liveliness*, and *strangeness* to already-controllable vibration engineering parameters which emotion controls could traverse. For designers and application developers, our results provide two automatable mappings and highlight their potential crosstalk.

Finally, we presented three example tuning interfaces and motivated future directions for research in haptic perception: (1) characterizing sensory parameters of vibrations (e.g., discontinuity, tempo) for tool development, and (2) validating and refining our mappings in lab-based perceptual studies or in-situ design-driven case studies.

Our results enable new interfaces for vibrotactile design and personalization, which in turn pave the way towards more expressive vibrotactile sensations and improved adoption and engagement by end-users.

References

1. Chan, A., MacLean, K., McGrenere, J.: Designing haptic icons to support collaborative turn-taking. Int. J. Hum.-Comput. Stud. (IJHCS) **66**(5), 333–355 (2008)
2. Ryu, J., Chun, J., Park, G., Choi, S., Han, S.H.: Vibrotactile feedback for information delivery in the vehicle. IEEE Trans. Haptics (ToH) **3**(2), 138–149 (2010). https://doi.org/10.1109/TOH.2010.1
3. Brunet, L., Megard, C., Paneels, S., Changeon, G., Lozada, J., Daniel, M.P., Darses, F.: Invitation to the voyage: The design of tactile metaphors to fulfill occasional travelers' needs in transportation networks. In: IEEE World Haptics Conference (WHC '13), pp. 259–264 (2013). https://doi.org/10.1109/WHC.2013.6548418
4. Israr, A., Zhao, S., Schwalje, K., Klatzky, R., Lehman, J.: Feel effects: enriching storytelling with haptic feedback. ACM Trans. Appl. Percep. (TAP) **11**, 11:1–11:17 (2014)
5. Schneider, O.S., MacLean, K.E.: Studying design process and example use with macaron, a web-based vibrotactile effect editor. In: Proceedings of IEEE Haptics Symposium (HAPTICS '16), pp. 52–58 (2016)
6. Tam, D., MacLean, K.E., McGrenere, J., Kuchenbecker, K.J.: The design and field observation of a haptic notification system for timing awareness during oral presentations. In: Proceedings of the ACM SIGCHI Conference on Human Factors in Computing Systems (CHI '13), pp. 1689–1698. ACM, New York (2013). https://doi.org/10.1145/2470654.2466223
7. Zhao, S., Schneider, O., Klatzky, R., Lehman, J., Israr, A.: Feelcraft: Crafting tactile experiences for media using a feel effect library. In: Proceedings of the Adjunct Publication of the 27th Annual ACM Symposium on User Interface Software and Technology (UIST '14), pp. 51–52. ACM, New York (2014). https://doi.org/10.1145/2658779.2659109
8. Lieberman, H., Paternò, F., Klann, M., Wulf, V.: End-user development: An emerging paradigm. In: End User Development, pp. 1–8. Springer, Berlin (2006)
9. Kwak, D.H., Clavio, G.E., Eagleman, A.N., Kim, K.T.: Exploring the antecedents and consequences of personalizing sport video game experiences. Sport Mark. Q. **19**(4), 217–225 (2010). http://ezproxy.library.ubc.ca/login?url=http://search.proquest.com/docview/851541557?accountid=14656. Copyright - Copyright Fitness Information Technology, A Division of ICPE West Virginia University Dec 2010; Document feature - Tables; Accessed 06 July 2012
10. Saul, G., Lau, M., Mitani, J., Igarashi, T.: Sketchchair: an all-in-one chair design system for end users. In: Proceedings of the Fifth ACM International Conference on Tangible, Embedded, and Embodied Interaction (TEI '11), pp. 73–80 (2011)
11. Seifi, H., Anthonypillai, C., MacLean, K.E.: End-user customization of affective tactile messages: A qualitative examination of tool parameters. In: Proceedings of IEEE Haptics Symposium (HAPTICS '14), pp. 251–256. IEEE (2014)
12. Seifi, H., Zhang, K., MacLean, K.E.: Vibviz: Organizing, visualizing and navigating vibration libraries. In: Proceedings of IEEE World Haptics Conference (WHC '15), pp. 254–259. IEEE (2015)
13. Adobe Systems, Inc.: Adobe photoshop. https://www.adobe.com/ca/products/photoshop.html. Accessed 23 Oct 2016
14. Evening, M.: The Adobe Photoshop Lightroom 5 Book: The Complete Guide for Photographers. Pearson Education, London (2013)
15. Facebook, Inc.: Instagram. https://www.instagram.com/?hl=en. Accessed 21 Oct 2016
16. Paterson, M.: The Senses of Touch: Haptics, Affects and Technologies. Berg (2007)
17. McGlone, F., Vallbo, A.B., Olausson, H., Loken, L., Wessberg, J.: Discriminative touch and emotional touch. Can. J. Exp. Psychol. **61**(3), 173 (2007)
18. Hertenstein, M.J., Weiss, S.J.: The Handbook of Touch: Neuroscience, Behavioral, and Health Perspectives. Springer Publishing Company, Berlin (2011)
19. Seifi, H., MacLean, K.E.: Exploiting haptic facets: users' sensemaking schemas as a path to design and personalization of experience. In: Submitted to International Journal of Human Computer Studies (IJHCS), Special issue on Multisensory HCI (2017)

20. Engineering Acoustics, Inc.: C2 tactor. https://www.eaiinfo.com/tactor-info/. Accessed 21 Mar 2017
21. MacLean, K.E., Schneider, O., Seifi, H.: Multisensory haptic interactions: understanding the sense and designing for it. In: The Handbook of Multimodal-Multisensor Interfaces. ACM Books (2017)
22. Blom, J.O., Monk, A.F.: Theory of personalization of appearance: Why users personalize their pcs and mobile phones. J. Hum.-Comput. Interact. **18**(3), 193–228 (2003). http://www.tandfonline.com/doi/abs/10.1207/S15327051HCI1803_1
23. McGrenere, J., Baecker, R.M., Booth, K.S.: A field evaluation of an adaptable two-interface design for feature-rich software. ACM Trans. Comput.-Hum. Interact. (TOCHI) **14**(1), 3 (2007). http://dl.acm.org/citation.cfm?id=1229858
24. Mackay, W.E.: Triggers and barriers to customizing software. In: Proceedings of ACM SIGCHI conference on Human Factors in Computing Systems (CHI '91), pp. 153–160 (1991). http://dl.acm.org/citation.cfm?id=108867
25. Marathe, S., Sundar, S.S.: What drives customization?: Control or identity? In: Proceedings of ACM SIGCHI Conference on Human Factors in Computing Systems (CHI '11), pp. 781–790 (2011). http://dl.acm.org/citation.cfm?id=1979056
26. Oh, U., Findlater, L.: The challenges and potential of end-user gesture customization. In: Proceedings of ACM SIGCHI Conference on Human Factors in Computing Systems (CHI '13), pp. 1129–1138 (2013). http://dl.acm.org/citation.cfm?id=2466145
27. Nurkka, P.: "Nobody other than me knows what i Want": customizing a sports watch. In: Kotz, P., Marsden, G., Lindgaard, G., Wesson, J., Winckler, M. (eds.) Proceedings of Human-Computer Interaction (INTERACT '13), vol. no. 8120 in Lecture Notes in Computer Science, pp. 384–402. Springer, Berlin, (2013). http://link.springer.com/chapter/10.1007/978-3-642-40498-6_30
28. Schneider, O.S., Israr, A., MacLean, K.E.: Tactile animation by direct manipulation of grid displays. In: Proceedings of the 28th Annual ACM Symposium on User Interface Software and Technology (UIST '15), pp. 21–30. ACM (2015)
29. Lee, J., Ryu, J., Choi, S.: Vibrotactile score: A score metaphor for designing vibrotactile patterns. In: Proceedings of IEEE World Haptics (WHC '09), pp. 302–307 (2009). https://doi.org/10.1109/WHC.2009.4810816
30. Google, Inc.: Google Play Music. https://play.google.com/music/listenow. Accessed 21 Oct 2016
31. Musicovery: Musicovery. http://musicovery.com/. Accessed 21 Oct 2016
32. MoodFuse: Moodfuse. https://moodfuse.com. Accessed 21 Oct 2016
33. Nik Software: Snapseed, on Google Play Store. https://itunes.apple.com/ca/app/snapseed/id439438619?mt=8. Accessed 21 Oct 2016
34. Propellerhead Software: Figure, on Google Play Store. https://www.propellerheads.se/products/figure/manual/introduction/. Accessed 21 Oct 2016
35. van Erp, J.B., Spapé, M.M.: Distilling the underlying dimensions of tactile melodies. Proc. Eurohaptics Conf. **2003**, 111–120 (2003)
36. Hoggan, E., Brewster, S.: Designing audio and tactile crossmodal icons for mobile devices. In: Proceedings of the 9th ACM International Conference on Multimodal Interfaces (ICMI '07), pp. 162–169. ACM (2007)
37. Brown, L.M., Brewster, S.A., Purchase, H.C.: Tactile crescendos and sforzandos: applying musical techniques to tactile icon design. In: CHI'06 Extended Abstracts on Human factors in Computing Systems (CHI EA '06), pp. 610–615. ACM (2006)
38. Mazzoni, D., Dannenberg, R.: Audacity Software. http://audacity.sourceforge.net/. Accessed 24 Jan 2015
39. SoundTouch: Soundtouch Algorithm (2016). http://www.surina.net/soundtouch/. Accessed 24 Sept 2016
40. Koskinen, E., Kaaresoja, T., Laitinen, P.: Feel-good touch: Finding the most pleasant tactile feedback for a mobile touch screen button. In: Proceedings of the 10th International Conference on Multimodal Interfaces (ICMI '08), pp. 297–304. ACM, New York (2008). http://doi.acm.org.ezproxy.library.ubc.ca/10.1145/1452392.1452453

41. Zheng, Y., Morrell, J.B.: Haptic actuator design parameters that influence affect and attention. In: Proceedings of IEEE Haptics Symposium (HAPTICS '12), pp. 463–470. IEEE (2012)
42. Schneider, O.S., MacLean, K.E.: Improvising design with a haptic instrument. In: Proceedings of IEEE Haptics Symposium (HAPTICS '14), pp. 327–332. IEEE (2014)
43. Yoo, Y., Yoo, T., Kong, J., Choi, S.: Emotional responses of tactile icons: Effects of amplitude, frequency, duration, and envelope. In: Proceedings of IEEE World Haptics Conference (WHC'15), pp. 235–240 (2015). https://doi.org/10.1109/WHC.2015.7177719
44. O'Sullivan, C., Chang, A.: An Activity Classification for Vibrotactile Phenomena, pp. 145–156. Springer, Berlin (2006). https://doi.org/10.1007/11821731_14
45. Schneider, O.S., Seifi, H., Kashani, S., Chun, M., MacLean, K.E.: Hapturk: crowdsourcing affective ratings of vibrotactile icons. In: Proceedings of the ACM SIGCHI Conference on Human Factors in Computing Systems (CHI '16), pp. 3248–3260 (2016)
46. Ternes, D., Maclean, K.E.: Designing large sets of haptic icons with rhythm. In: Haptics: Perception, Devices and Scenarios, pp. 199–208. Springer, Berlin (2008)
47. Wobbrock, J.O., Findlater, L., Gergle, D., Higgins, J.J.: The aligned rank transform for nonparametric factorial analyses using only anova procedures. In: Proceedings of the SIGCHI Conference on Human Factors in Computing Systems (CHI '11), pp. 143–146. ACM (2011)
48. Conover, W.J., Iman, R.L.: Rank transformations as a bridge between parametric and nonparametric statistics. Am. Stat. 35(3), 124–129 (1981). http://amstat.tandfonline.com/doi/abs/10.1080/00031305.1981.10479327

Chapter 8
Conclusion and Future Directions

Abstract *We envision, for haptic personalization, a suite of tools that are unified by one underlying conceptual model and can be effectively incorporated into users' workflows with various applications.* Our vision is analogous to what we have for color personalization: A simple, yet powerful, suite of tools (color swatches, color gamut, sliders) built on the color theory (e.g., Munsell color system), and seamlessly integrated in a variety of applications (e.g., Microsoft Office, Adobe Creative Suite) for a wide range of users. Below, we first discuss the contributions of this book towards this vision, divided by the three main themes that we presented in Chap. 1, then we outline the next steps for which the research described here has exposed a need.

8.1 Personalization Mechanisms

Identifying a suite of tools: Our first study on personalization mechanisms suggests the need for a *set* of personalization mechanisms (i.e., tools), rather than a single one. In our lab-based study, users varied in the personalization mechanism they preferred, weighing "design effort", "sense of control", and "fun" differently. To inform tool design, we outlined the design space for personalization mechanisms and proposed three promising candidate mechanisms (*choosing*, *tuning*, and *chaining*) for the personalization tool suite. This study, however, mainly examined the concepts of these mechanisms, without providing any guidelines for realizing them as tools.

Developing the mechanisms: *Can these mechanisms be developed into tools? What would those tools look like?* With *VibViz* (Chap. 4), we built on the principles from information visualization and library sciences to devise a set of guidelines for a *choosing* interface. In Chap. 7, we verified the feasibility of developing automatic *emotion* controls for *tuning* vibrations and presented three example interfaces for this mechanism (Fig. 8.1).

© Springer Nature Switzerland AG 2019   163
H. Seifi, *Personalizing Haptics*, Springer Series on Touch
and Haptic Systems, https://doi.org/10.1007/978-3-030-11379-7_8

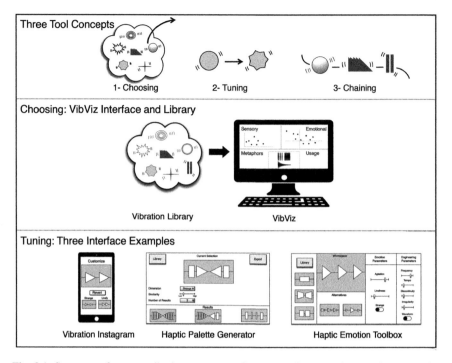

Fig. 8.1 Summary of our contributions to *personalization mechanisms*: three tool concepts for personalization (Chap. 3) and development of *choosing* (Chap. 4) and *tuning* (Chap. 7) concepts

Reflecting on alternative designs: While our prototypes were developed for a particular platform, their underlying mechanisms are platform independent. In this book, we focused on developing the mechanisms and chose the prototyping platforms (e.g., device, programming technology) based on an anticipated personalization workflow given a tool. For example, *VibViz* was designed for devices with a relatively large screen size (e.g., a desktop or laptop computer) where users would want to explore a wide range of pre-designed vibrations before *choosing* one. Similarly, the *Haptic Palette Generator* and *Emotion Toolbox* are designed for stationary use cases. In contrast, the *Vibration Instagram* prototype was designed for a phone interface where users can apply quick fixes on the go.

However, this association is not rigid. Our designs can be revised and adapted to alternative platforms to accommodate diverse use cases and user preferences. An increasing portion of the users spend most of their time on mobile devices (e.g., smartphones and smartwatches). Therefore, desktop applications, designed for everyday use, typically have an accompanying mobile application. Our prototypes can be redesigned to accommodate smaller screen sizes. For example, a mobile version of *VibViz* can present one facet view at a time (e.g., Sensory and Emotional View) while allowing users to switch to the other views (e.g., tabs in an interface) when needed. Search filters can be presented with common interface widgets such as a navigation drawer. The new design would have reduced functionality compared to the desktop version (e.g., users cannot easily crosslink vibrations on different views) but could enable quick selections. Alternatively, a small screen size can lead to a very different design. For example, a *choosing* interface can present a subset of the vibrations in the library based on the users' preference and interaction history. Such interfaces would benefit from future research on adaptive interfaces and recommender systems for haptic sensations.

8.2 Facets as an Underlying Model for Personalization Tools

Supporting users' mental model: Facets offer a unifying conceptual model for the haptic personalization tools. In this respect, their primary advantage is their match with users' cognitive structure(s). Our five proposed facets are derived from people's descriptions for haptic sensations and encapsulate their multiple and overlapping sense-making schemas for haptics. *VibViz* showed that these facets, even in their primary form as a flat list of attributes, can enable design of powerful tools for end-users. In *VibViz*, several linked views of the vibration library supported an individual's varying criteria in different usage contexts as well as differences among the users in cognition and preference.

Informing design practices: Informing design beyond a tool's interface requires a concise picture of the facets as well as a path from the facets to *sensation* and *engineering parameters* available in haptic authoring tools and display hardware. In Chap. 5, we derived a set of semantic dimensions for the facets, thereby structured their large list of attributes into a succinct set. Further, we linked a path from the *emotion*, *metaphor*, and *usage* facets to the *sensation* facet.

Linking the facets to engineering parameters: In Chap. 7, we verified that the *tuning* mechanism can be built on our evolved understanding of the facets, namely their semantic dimensions and interlinkages. Focusing on *emotion* controls, we present a path from the three *emotion* dimensions to *engineering parameters* of a specific actuator (C2 tactor), using the linkages between the *emotion* and *sensation* facets as a middle step.

Reporting individual differences: In developing end-user tools and/or rich sensations, designers must note variations around an aggregated model. Thus, we also present an in-depth analysis of individual differences in the facets and their attributes (Fig. 8.2).

Reflecting on the facets: Facets and their interlinkages are an evolving concept and our work is a first pragmatic characterization of them. Notably, our proposed facets can overlap, with some attributes being applicable to more than one facet. For example, "alarm" can belong to both the *usage example* and *metaphor* facets. Similarly, "energetic" can be included in the *sensation* or *emotion* facets. These instances question the idea of rigid boundaries for the facets, and suggest an evolving and

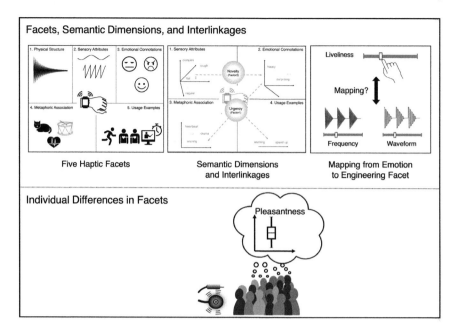

Fig. 8.2 Summary of our contributions to *affective haptics*: five haptic facets (Chap. 4), their semantic dimensions and linkages (Chap. 5), and quantification of individual differences in affect (Chaps. 2, 5, and 6)

flexible characterization that can be revised, shifted, or combined as our understanding of the domain evolves and depending on the use case.

Examples of these revisions and shifts can be seen in this books. Initially, we defined the *physical* facet to encapsulate all measurable properties of vibrations including *energy* and *tempo* attributes. As our understanding of the facets evolved, we moved these two attributes to the *sensation* facet (since they cannot be measured objectively, at least not yet) and revised the *physical* facet to include the *engineering* parameters (e.g., duration, frequency, waveform). As another example, in designing the *VibViz* interface, we combined the *sensation* and *emotion* facets in one view for a more effective access to the dataset. Finally, our evolving understanding of the facets is reflected in our naming; we started with calling the schemas "taxonomies" and later switched to "facets" as it denotes a flexible structure that can combine a mix of attributes with different characteristics (e.g., numerical ratings, words) in a flat and/or hierarchical structure depending on the known semantic linkages between the attributes. Future work can further refine these facets and/or add new unexplored ones.

8.3   Large Scale Evaluation for Theory and Tool Development

Devising methodologies for haptic studies: In developing theories and tools for affective haptic design, the haptic community needs to study large and diverse groups of users, yet haptic methodologies rarely scale. We faced the need for scaling our studies during this research and devised new evaluation methodologies that work around existing practical limitations in the field. Our two methodological contributions address the same problem but have unique elements which makes them suitable for different contexts. The two-stage methodology enables lab-based studies and integrates experts' evaluation with data from lay users. In contrast, with the crowdsourcing methodology, researchers can collect fast and inexpensive data from a large group but within an error threshold (Fig. 8.3).

Reflecting on the long-term value of our methodologies: Fast technological progress and proliferation of the technology can resolve some of the existing problems in accessing the crowds and expedite crowdsourcing in the haptic community. An important question is: *would widespread access to haptic hardware eliminate the need for haptic proxies or in-lab studies?*

We believe the answer is no. Haptics still has a long journey ahead to achieve the full range of natural sensations. New technologies are being developed everyday, adding to the expressive range of existing hardware and/or enabling newly programmable sensations. These new technologies usually go under several rounds of

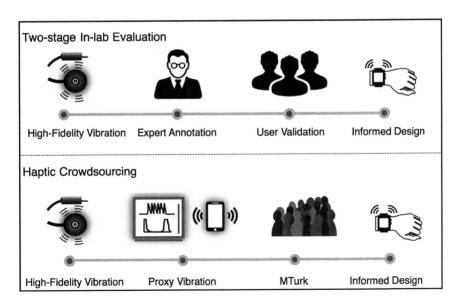

Fig. 8.3 Summary of our contributions to *large scale evaluation*: two methodologies for evaluating haptics in lab (Chap. 5) and online (Chap. 6)

research, design, and evaluation to mature and pass cost-value trade-offs of business units. Thus, there will always be a gap between the new technologies tested in research labs and the ones available to everyday users, necessitating the use of proxies or in-lab evaluations.

Further, the haptic industry plays an important role in the progress of the field. Often, companies are not willing to expose their design(s) on online platforms, yet are interested in efficient evaluation methods. Thus, using experts for evaluation is highly desirable as an alternative discount evaluation method when access to the crowds is limited. We hope our contributions facilitate a range of haptic perception studies and inspire new methodologies that further expand designers' and researchers' evaluation toolkit.

8.4   Future Work

8.4.1   Incorporating Personalization in Users' Workflow

A key aspect of our vision was effective integration of the personalization tools in users' natural workflow with an application. Otherwise, personalization tools will either be abandoned or at best adopted by a niche group of power users. This requirement further highlights open questions about the users' personalization practices and workflows. Specifically:

What workflows and scenarios can best support the users? Where do our tool approaches lie in the personalization process? What are the other requirements besides effective tools? Given effective tools and personalization workflows, do people personalize haptic signals for their everyday devices?

These questions cannot be effectively addressed until a user base has built up a body of experience and needs around this new technology. Currently, a majority of users have little exposure to the range of possible vibrotactile signals beyond the dull notification buzz on their cellphones and are unsure of application possibilities of haptics. Thus, they cannot reliably judge their interest in personalization, nor can they reflect on their personalization workflows. Our work focuses on developing effective personalization mechanism, the groundwork needed to tackle the above questions. Further, our lab-based studies of these mechanisms and past haptic field studies provide evidence for the importance of personalization, thereby motivate further research on the above questions.

Rigorous answer(s) to these questions require a series of studies triangulating various research methods; including in-situ studies of haptic applications and personalization practices, small-scale longitudinal qualitative studies, and large scale deployment of personalization tools. Results can inform future haptic tools and facilitate integration of haptic personalization in end-user devices and applications.

8.4.2 Expanding the Mechanisms and the Underlying Model

What are other effective mechanisms in the personalization design space? Can the facets inform those mechanisms? What are alternative underlying models for personalization tools?

Emerging paradigms for haptic sketching may inform design of new personalization mechanisms. With recently developed haptic authoring tools, people can rapidly create a sensation by demonstrating its properties in a more accessible modality (e.g., by drawing, vocalizing, tapping). Apple's iOS has a simple interface where users can tap a pattern to create a custom vibration. With mHive, users can create a vibrotactile sensation by drawing a path on a tablet touchscreen [1]. Voodle is an example system, developed in our lab, where users can control movements of a 1-DoF robot with their voice in real time [2].

While these interfaces are effective for design [1, 2], they need to be revised and adapted for personalization. In their current form, these interfaces are too open-ended for most users, as evidenced by the negative comments on the iOS's tapping interface (Chap. 3). One possibility is using a mixed-initiative approach where the users sketch a pattern (e.g., by drawing, vocalizing, or tapping) and the system renders and refines it into a plausible sequence based on a set of perceptual rules. Alternatively, the systems can recommend a set of patterns, from a large repository, based on input sequences sketched by the users. Future studies can investigate these and other plausible mechanisms for personalization to complement the suite of tools available to the users.

In developing such new mechanisms, researchers can determine the utility of the facets as an underlying conceptual model and/or propose alternative models for personalization.

8.5 Final Remarks

The study of haptic design and in particular, the affective aspect of designed haptic experiences, was largely ignored until very recently. Premier haptic conferences, namely Haptic Symposium and World Haptics, were mainly focused on hardware development and users' tactile and kinesthetic abilities, while main HCI conferences such as CHI sometimes published studies that were not considered novel among haptics experts. Further, for the first decade of consumer-level haptic devices, the quality of the haptic experience offered to end-users was very low. Phones included low-fidelity actuators, resulting in the users' low opinion of haptics.

But the situation is rapidly changing. Open haptics communities are being formed, where the goal is to share design contributions widely, discuss avenues for further progress, and eliminate several cases of "reinventing the wheel" in hardware development or perceptual studies. The haptic community is increasingly recognizing the importance of HCI and design. *VibViz* (Chap. 4) and *Macaron* [3] were nominated for best demo awards, in World Haptics 2015 and Haptic Symposium 2016 respectively, for their contribution to affect and design. Both tools also received great attention from the haptic industry and academia. In particular, Immersion Inc., the world's largest haptic company, contacted our group to utilize the design ideas from *VibViz* in their internal tools. They were interested in providing their designers with a unified interface for accessing their several haptic libraries efficiently. Apple has recently integrated a high-fidelity voice coil haptic engine in their smartwatch, pioneering the change in future devices and suggesting exciting possibilities for engaging the crowds.

The content of this book has played a pioneering role in the above changes and specifically in the areas of affective haptics and design. Here we tackled an unexplored area of haptics: end-user personalization. We provided a theoretical grounding for personalization tools (facets and personalization mechanisms) and prototyped example interfaces (*VibViz*, and three *tuning* interfaces) to showcase tool design possibilities. Further, we pushed the boundaries of haptic evaluation, investigating crowdsourcing and use of haptics experts. We hope this work sparks future research in haptic design, aesthetics, and personalization and ultimately contributes to fun, informative, and satisfying haptic experiences for all individuals.

References

1. Schneider, O.S., MacLean, K.E.: Improvising design with a haptic instrument. In: Proceedings of IEEE Haptics Symposium (HAPTICS '14), pp. 327–332. IEEE (2014)
2. Schneider, O., Marino, D., Bucci, P., MacLean, K.E.: Voodle: vocal doodling for affective robot interaction. In: Proceedings of the Designing Interactive Systems, DIS '17 (2017)
3. Schneider, O.S., MacLean, K.E.: Studying design process and example use with macaron, a web-based vibrotactile effect editor. In: Proceedings of IEEE Haptics Symposium (HAPTICS '16), pp. 52–58 (2016)

Chapter 9
Supplemental Materials for Chapter 5

Abstract Here, we present additional data and analysis on the four haptic facets presented in Chap. 5. Specifically, here we present the list of tags in all four facets with their disagreement scores (Tables 9.1, 9.2, 9.3 and 9.4) as well as percentage of tags removed by different subsets of participants (Table 9.5). In addition, we report correlation between the five Likert-scale ratings in our user study (Table 9.6) and illustrate results of multidimensional scaling (MDS) analysis on the tags for the four facets in Figs. 9.1, 9.2, 9.3, and 9.4. These MDS results complement the MDS analysis reported in Sect. 5.5.1. Finally, we provide quantitative data on the individual differences in tag descriptions for all the vibrations (Fig. 9.6) and co-occurrence rates between the tags in the sensory facet and the other three facets (Figs. 9.7, 9.8 and 9.9).

9.1 List of Tags and Their Disagreement Values

In this section, we present the full list of tags collected for the four vibration facets along with their disagreement scores (Tables 9.1, 9.2, 9.3 and 9.4).

© Springer Nature Switzerland AG 2019   171
H. Seifi, *Personalizing Haptics*, Springer Series on Touch
and Haptic Systems, https://doi.org/10.1007/978-3-030-11379-7_9

Table 9.1 Sensation$_f$ tags and disagreement scores

Index	Tag	Disagreement score
1	Short	0.08
2	Smooth transition	0.09
3	Irregular	0.11
4	Pointy	0.11
5	Ramping up	0.12
6	Grainy	0.12
7	Long	0.13
8	Simple	0.17
9	Firm	0.17
10	Rough	0.17
11	Wavy	0.17
12	Continuous	0.17
13	Discontinuous	0.17
14	Bumpy	0.17
15	Dynamic	0.2
16	Regular	0.2
17	Spiky	0.21
18	Soft	0.22
19	Springy	0.22
20	Smooth	0.22
21	Ramping down	0.24
22	Complex	0.28
23	Flat	0.28
24	Ticklish	0.31

Table 9.2 Emotion$_f$ tags and disagreement scores

Index	Tag	Disagreement score
1	Rhythmic	0.14
2	Attention-getting	0.16
3	Agitating	0.18
4	Unique	0.18
5	Energetic	0.18
6	Mechanical	0.19
7	Familiar	0.2
8	Surprising	0.21
9	Urgent	0.22
10	Natural	0.22
11	Strange	0.23

(continued)

Table 9.2   (continued)

Index	Tag	Disagreement score
12	Predictable	0.24
13	Uncomfortable	0.25
14	Lively	0.25
15	Calm	0.26
16	Interesting	0.26
17	Annoying	0.27
18	Comfortable	0.27
19	Pleasant	0.31
20	Happy	0.31
21	Angry	0.32
22	Boring	0.32
23	Creepy	0.32
24	Sad	0.34
25	Fear	0.36
26	Funny	0.36

Table 9.3   Metaphor$_f$ tags and disagreement scores

Index	Tag	Disagreement score
1	Dancing	0.11
2	Pulsing	0.11
3	Getting close	0.11
4	Cymbal	0.11
5	Alarm	0.15
6	Phone	0.15
7	Morse code	0.16
8	Heart beat	0.17
9	SOS	0.18
10	Buzz	0.18
11	Engine	0.19
12	Sliding	0.2
13	Tapping	0.21
14	Game	0.22
15	Going away	0.22
16	Shaking	0.22
17	A door closing	0.22
18	Stopping	0.22

(continued)

Table 9.3 (continued)

Index	Tag	Disagreement score
19	Growl	0.22
20	Frogs	0.22
21	Poking	0.23
22	Coming or going	0.23
23	Beep	0.24
24	Horn	0.25
25	Jumping	0.25
26	Snoring	0.27
27	Riding	0.28
28	Clock	0.28
29	Drums	0.28
30	Breathing	0.3
31	Electric shock	0.3
32	Musical instruments	0.3
33	Nature	0.31
34	Bell	0.31
35	Gun	0.31
36	Pawing	0.31
37	Celebration	0.31
38	Walking	0.33
39	Echo	0.33
40	Explosion	0.33
41	Chainsaw	0.33
42	Animal	0.34
43	A spring	0.44
44	Footsteps	0.44
45	A story	0.44

Table 9.4 Usage $_f$ tags and disagreement scores

Index	Tag	Disagreement score
1	Alarm	0.21
2	Halfway	0.21
3	Reminder	0.22
4	Warning	0.22
5	Running out of time	0.23
6	Confirmation	0.23
7	Speed up	0.24

(continued)

Table 9.4 (continued)

Index	Tag	Disagreement score
8	Overtime	0.24
9	Slow down	0.25
10	Interval/rep	0.25
11	Above intended threshold	0.26
12	Resume	0.26
13	One minute left	0.27
14	Finish	0.27
15	Incoming message	0.28
16	Congratulations	0.28
17	Get ready	0.3
18	Milestone	0.3
19	Encouragement	0.3
20	Battery low	0.3
21	Pause	0.3
22	Warm up	0.31
23	Cool down	0.31
24	Below intended threshold	0.33
25	Start	0.36

9.2 Tag Removal Summary

Table 9.5 summarizes the percentage of tags removed by the lay users in the validation study.

Table 9.5 Percentage of tags removed by normal users. Each row represents the percentages of tags that are removed by at least x people (x = 1 for $\geq$ 1 label) in each facet (columns)

Number of participants removing a tag	Sensation$_f$ (%)	Emotion$_f$ (%)	Metaphor$_f$ (%)	Usage$_f$ (%)
≥ 1	79	88	87	92
≥ 2	51	69	67	74
≥ 3	27	46	43	53
≥ 4	14	28	24	31
≥ 5	8	14	10	15
≥ 6	4	6	3	7
≥ 7	2	2	1	2
≥ 8	1	0	0	0
≥ 9	1	0	0	0

9.3 Rating Correlations

The following table summarizes results of the Pearson correlation on the five rating
scales (Table 9.6).

Table 9.6 Results of Pearson correlation on the five rating scales. The correlation is applied on all
valid participants' ratings for the 120 vibrations

	Energy$_d$	Tempo$_d$	Roughness$_d$	Pleasantness$_d$	Arousal$_d$
Energy$_d$	1.00	0.48	0.74	−0.46	0.92
Tempo$_d$	0.48	1.00	0.52	−0.22	0.56
Roughness$_d$	0.74	0.52	1.00	−0.61	0.79
Pleasantness$_d$	−0.46	−0.22	−0.61	1.00	−0.53
Arousal$_d$	0.92	0.56	0.79	−0.53	1.00

9.4 Multidimensional Scaling Graphs on Tag Distances

Figures 9.1, 9.2, 9.3, 9.4 and 9.5 depict results of our MDS analysis on tag distances
in the four vibration facets.

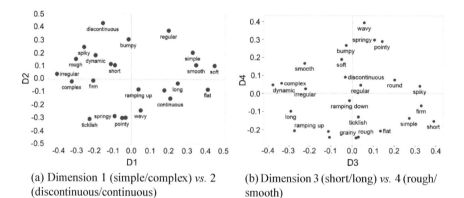

(a) Dimension 1 (simple/complex) *vs.* 2   (b) Dimension 3 (short/long) *vs.* 4 (rough/
(discontinuous/continuous)   smooth)

Fig. 9.1 Spatial configuration of the tags for the sensation$_f$ facet confirms the four identified
dimensions in Chap. 5. Specifically, contrasting tags according to each dimension are well-separated,
and the semantically-related tags are close together along each dimension

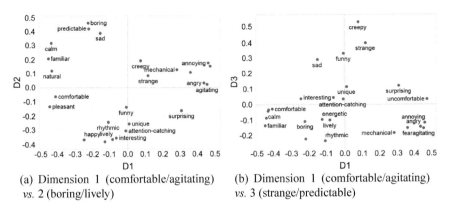

(a) Dimension 1 (comfortable/agitating) vs. 2 (boring/lively)

(b) Dimension 1 (comfortable/agitating) vs. 3 (strange/predictable)

Fig. 9.2 Spatial configuration of the tags for the emotion$_f$ facet confirms the three identified dimensions in Chap. 5 and supports convergent and discriminant validity

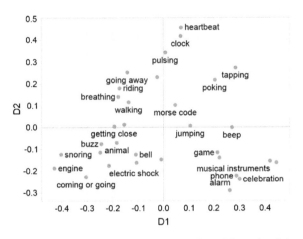

Fig. 9.3 Spatial configurations of tags for the metaphor$_f$ facet. Dimension 1 (on-off—ongoing$_d$) versus dimension 2 (natural—mechanical$_d$). Semantically-related tags, according to a dimension, are close along the dimension (e.g., drums, celebration, alarm) and contrasting tags are far from each other (e.g., heartbeat versus engine or alarm). This definition partially explains a few tags, such as clock (among the natural, calm sensations) and snoring (with mechanical, annoying and ongoing tags)

Fig. 9.4 Spatial
configurations of tags for the
usage$_f$ facets. Dimension 1
(urgent/awareness
notifications). Dimension 2
is not used in our analysis.
Along Dimension 1, tags
have increasing urgency and
attention demand from left to
right, supporting convergent
and discriminant validity for
the semantics of the
dimension

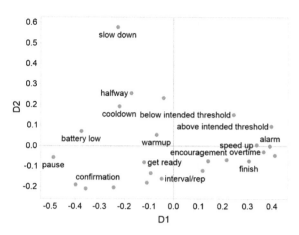

	Energy	Tempo	Roughness	Valence	Arousal	Sensation	Emotion	Metaphor	Usage
v-09-10-3-56	0.44	0.62	0.59	0.91	0.79	0.02	0.39	0.33	0.29
v-09-10-4-25	0.35	0.59	1.26	0.96	0.81	0.3	0.43	0.07	0.28
v-09-10-6-46	0.4	0.4	0.67	0.91	0.84	0.27	0.2	0.44	0.37
v-10-28-7-35	0.99	1.49	1.16	1.19	0.53	0.16	0.31	0.41	0.4
v-09-12-1-0	0.46	1.39	0.86	1.07	0.83	0.27	0.41	0.24	0.25
v-09-09-8-11	1.08	0.54	0.59	0.63	0.5	0.15	0.22	0.22	0.27
v-09-09-8-20	0.92	0.65	1.1	0.91	0.76	0.22	0.17	0.19	0.3
v-09-09-8-20-cpg	0.85	0.36	0.85	0.76	0.69	0.22	0.14	0.19	0.21
v-09-09-8-24	0.89	1.1	1.11	1.38	0.62	0.2	0.36	0	0.33
v-09-10-11-55	0.47	0.96	1.1	1.01	0.83	0.09	0.17	0.22	0.23
v-09-10-12-11	0.57	0.53	0.28	0.99	0.59	0.15	0.09	0.11	0.06
v-09-10-12-13	0.28	0.58	0.65	1.18	0.38	0.09	0.26	0.14	0.26
v-09-10-12-16	0.4	1.11	0.79	1.33	0.44	0.17	0.27	0.44	0.24
v-09-10-12-2	0.79	0.36	0.52	0.86	0.28	0.16	0.2	0.22	0.22
v-09-10-12-6	0.77	0.62	0.46	0.93	1.01	0.03	0.17	0.16	0.27
v-09-10-12-9	0.53	0.5	0.66	0.88	0.18	0.21	0.26	0.27	0.28
v-09-10-12-9-cpg	0.68	0.33	0.73	1.11	0.66	0.24	0.19	0.38	0.39
v-09-10-3-52	0.14	0.26	0.57	0.83	0.64	0.2	0.28	0.24	0.22
v-09-10-4-2	0.6	0	0.71	1.26	0.66	0.25	0.37	0.29	0.32
v-09-10-4-20	0.91	1.21	0.98	1.49	0.66	0.37	0.29	0.27	0.37
v-09-10-4-23	0.49	0.59	1.11	1.11	0.79	0.16	0.25	0.31	0.31
v-09-10-4-6	0.72	0.63	0.88	1.02	0.98	0.07	0.15	0.15	0.13
v-09-10-6-16	0.86	0.35	0.59	1.19	0.91	0.3	0.3	0.26	0.4
v-09-10-6-22	0.47	0.72	0.71	0.93	1.16	0.26	0.29	0.17	0.11
v-09-10-6-27	0.14	0	0.78	0.92	0.14	0.13	0.22	0.18	0.13
v-09-10-6-38	0.35	0.68	0.99	1.07	0.33	0.13	0.4	0.26	0.35
v-09-10-6-43	0.53	0.82	0.83	0.88	0.71	0.19	0.18	0.32	0.29
v-09-10-6-5	0.2	0.35	0.44	0.94	0.74	0.13	0.38	0.2	0.33
v-09-10-6-59	0.64	1.02	0.63	0.64	0.53	0.14	0.19	0.22	0.22
v-09-10-7-34	1.09	1.04	1.1	1.33	0.89	0.11	0.3	0.19	0.31
v-09-10-7-36	0.53	0.66	1.02	0.88	0.46	0.17	0.25	0.11	0.26
v-09-10-7-9	0.45	0.53	1.05	1.08	1.11	0.18	0.21	0.22	0.19
v-09-10-8-5	0.53	0.69	1.14	0.84	0.91	0.22	0.39	0.33	0.38
v-09-10-8-7	1.01	0.47	0.65	0.8	0.77	0.13	0.04	0.04	0.24
v-09-10-8-7-cpg	0.72	0.47	0.66	0.9	0.92	0.11	0.17	0.07	0.2
v-09-11-3-12	0.69	0.53	0.88	1.29	0.55	0.3	0.31	0.11	0.33
v-09-11-3-16	0.75	0.28	0.82	0.93	0.63	0.02	0.24	0.08	0.19
v-09-11-3-19	0.4	0.78	1.06	1.27	0.17	0.28	0.29	0.3	0.26
v-09-11-3-21	0.49	0.49	0.4	0.62	0.67	0.13	0.33	0.38	0.3
v-09-11-3-24	0.89	0.4	0.89	0.62	0.59	0.2	0.21	0.41	0.27
v-09-11-3-4	0.5	0.5	0.44	0.47	0.42	0.24	0.26	0.28	0.3
v-09-11-3-43	0.5	0.5	0.74	1.06	0.58	0.35	0.31	0.29	0.44
v-09-11-3-50	1.02	0.99	0.48	0.99	1.01	0.22	0.36	0.22	0.36
v-09-11-3-54	0.72	0.56	1.11	0.63	0.58	0.22	0.12	0.25	0.18
v-09-11-3-56	0.89	0.59	0.89	1.04	0.96	0.13	0.25	0.47	0.36
v-09-11-3-8	0.42	0.33	0.56	0.85	0.58	0.18	0.33	0.42	0.27
v-09-11-4-1	0.62	0.49	0.89	0.44	0.44	0.24	0.3	0.25	0.38
v-09-11-4-12	0.88	0.42	0.44	0.58	0.94	0.24	0.33	0.33	0.22
v-09-11-4-22	0.2	1.23	0.52	0.81	0.67	0.21	0.25	0.41	0.33
v-09-11-4-3	0.31	0.5	1.11	0.67	0.67	0.09	0.14	0.29	0.24
v-09-11-4-41	1.14	0.79	1.42	0.93	0.76	0.32	0.35	0.35	0.36
v-09-11-4-41-cpg	1.11	0.99	1.54	1.22	1.21	0.37	0.3	0.37	0.42
v-09-11-4-54	1.19	0.46	0.76	1.36	0.17	0.2	0.22	0.21	0.2
v-09-11-4-8	1.22	0.45	0.68	1.31	0.74	0.19	0.21	0	0.46
v-09-12-1-19	0.69	0.49	0.74	0.74	0.67	0.33	0.29	0.49	0.33
v-09-12-1-23	0.71	0.65	1.14	1.17	1.17	0.13	0.11	0.07	0.19
v-09-12-1-29	0.89	0.85	1.23	0.9	1.03	0.17	0.27	0.15	0.17
v-09-12-1-39	1.29	1.52	1.7	1.22	1.07	0.2	0.17	0.38	0.35
v-09-12-1-48	0.5	0.39	0.65	1.14	0.43	0.19	0.38	0.28	0.23
v-09-12-1-53	0.81	0.44	1.01	1.11	0.81	0.15	0.24	0.53	0.32
v-09-12-2-17	0.81	1.11	1.43	1.19	1.26	0.17	0.17	0.31	0.35
v-09-12-2-20	0.54	0.83	1.17	1.24	1.11	0.14	0.32	0.17	0.2
v-09-12-2-23	0.92	0.58	1.02	0.89	0.75	0.13	0.3	0.16	0.23
v-09-12-2-40	0.58	0.89	0.45	1.22	1.4	0.2	0.42	0.41	0.48
v-09-12-8-10	0	1.22	1.07	1.98	0.53	0.22	0.37	0.28	0.38

Fig. 9.5 Vibration disagreement scores for the five rating scales and the four facets. High color saturation denotes high disagreement scores (part A: vibrations 1–60 in the VibViz library)

9.5   Individual Differences in Vibrations

The following table present disagreement scores calculated for the 120 vibrations in the *VibViz* library (Fig. 9.6).

	Energy	Tempo	Roughness	Valence	Arousal	Sensation	Emotion	Metaphor	Usage
v-09-12-8-13	1.18	1.04	1.14	0.8	1.15	0.04	0.25	0.14	0.32
v-09-12-8-21	0.93	1.03	1.01	0.86	0.84	0.09	0.11	0.17	0.19
v-09-12-8-27	0.62	1.19	1.26	0.96	0.69	0.06	0.13	0.33	0.39
v-09-12-8-30	0.17	1.21	0.93	1.11	0.6	0.2	0.41	0.41	0.39
v-09-12-8-32	0.74	0.57	0.96	1.53	1.04	0.27	0.3	0.44	0.34
v-09-16-1-43	0.67	0.62	0.96	1.04	0.69	0.26	0.28	0.22	0.28
v-09-16-1-43-cpy	0.69	1.01	1.19	1.04	0.62	0.24	0.33	0.33	0.24
v-09-16-1-56	0.71	0.5	0.91	0.8	0.53	0.17	0.23	0.13	0.19
v-09-18-1-55	0.56	0.75	1	1.25	0.5	0.27	0.27	0.11	0.32
v-09-18-2-7	0.46	0.71	0.81	1.36	0.3	0.13	0.23	0.28	0.37
v-09-18-4-12	1.22	0.83	0.79	0.66	0.66	0.16	0.26	0.25	0.38
v-09-18-4-15	0.94	0.2	0.59	0.84	0.52	0.14	0.26	0.26	0.33
v-09-18-4-16	1.12	0.63	0.74	1.21	0.96	0.17	0.43	0.24	0.32
v-09-18-4-18	0.54	0.59	0.77	1.05	0.57	0.06	0.18	0.14	0.15
v-09-18-4-22	1.43	1.11	1.13	0.68	0.83	0.23	0.31	0.28	0.33
v-09-18-4-56	1.75	1.62	0.94	1.29	1.36	0.32	0.39	0.38	0.26
v-09-23-6-24	1.7	0.96	1.41	1.26	0.89	0.22	0.29	0.3	0.35
v-09-23-6-24-cpy	1.33	0.99	1.16	1.58	0.59	0.28	0.25	0.26	0.4
v-09-26-1-39	1.53	1.23	0.99	1.23	0.91	0.14	0.19	0.11	0.17
v-10-09-1-1	0.33	0.73	0.98	1.09	0.88	0.24	0.36	0.26	0.44
v-10-09-1-11	0.44	0.89	0.67	0.74	0.89	0.19	0.19	0.44	0.36
v-10-09-1-12	1	0.82	0.58	0.92	1.31	0.15	0.33	0.22	0.22
v-10-09-1-14	0.9	0.52	0.54	0.93	0.66	0.07	0.05	0.08	0.27
v-10-09-1-16	0.75	0.83	0.72	0.77	0.63	0.15	0.22	0	0.28
v-10-09-1-16-cpy	0.73	0.76	1.04	1.02	0.54	0.15	0.25	0	0.24
v-10-09-1-20	0.83	0.78	0.67	1.26	0.61	0.17	0.35	0.4	0.36
v-10-09-1-8	1.39	1.11	0.88	1.25	1.17	0.3	0.33	0.22	0.24
v-10-09-1-8-cpy	1.17	1.22	0.72	1.32	1.08	0.21	0.37	0.33	0.28
v-10-09-5-0	0.4	0.2	0.79	0.81	0.81	0.22	0.17	0.26	0.26
v-10-09-5-2	0.69	0.46	0.45	1.42	0.66	0.22	0.22	0.29	0.34
v-10-09-5-4	0.66	0.96	0.69	1.11	0.45	0.36	0.35	0.38	0.31
v-10-09-5-7	0.44	0.28	0.58	1.17	0.49	0.24	0.24	0.25	0.27
v-10-09-5-7-cpy	0.28	0.28	0.44	0.92	0.49	0.22	0.2	0.36	0.32
v-10-10-1-10	0.6	0.45	0.51	0.66	0.99	0.31	0.26	0.44	0.36
v-10-10-1-10-cpy	1.12	1.06	0.84	1.19	0.45	0.33	0.36	0.5	0.33
v-10-10-1-18	0.88	0.48	1.39	1.22	0.58	0.33	0.35	0.29	0.38
v-10-10-1-21	0.43	0.67	0.39	1.05	0.47	0.11	0.17	0.11	0.14
v-10-10-1-5	0.59	0.8	0.93	1.11	0.63	0.11	0.31	0.06	0.18
v-10-18-11-11	0.15	0.38	0.88	0.69	0.15	0.24	0.19	0.14	0.34
v-10-21-2-48	0.85	0.58	0.65	1.15	0.66	0.11	0.21	0.2	0.15
v-10-21-3-11	0.14	0.28	0.47	0.99	0.28	0.13	0.24	0.33	0.24
v-10-21-3-17	0.15	0.67	0.68	0.92	0.15	0.14	0.28	0.19	0.3
v-10-21-3-2	0.74	0.4	1.19	0.67	0.59	0.16	0.33	0.48	0.28
v-10-21-3-21	0.89	1.31	0.55	0.78	1.12	0.25	0.33	0.46	0.32
v-10-21-3-30	1.17	0.67	1.06	1.22	0.67	0.11	0.39	0.29	0.37
v-10-21-3-33	1.35	0.15	1.33	1.75	0.96	0.13	0.35	0.17	0.15
v-10-21-3-39	0.89	0.43	0.53	0.58	0.47	0.04	0.21	0.08	0.24
v-10-21-3-4	0.64	0.26	1.43	1.08	0.47	0.04	0.17	0.12	0.16
v-10-21-3-45	0.14	0.28	0.39	0.8	0.28	0.14	0.13	0.13	0.13
v-10-21-3-45-cpy	0.47	0.31	0.57	0.67	0.36	0.22	0.18	0.17	0.1
v-10-21-3-7	1.08	0.38	1.03	1.03	0.92	0.19	0.24	0.19	0.23
v-10-23-1-10	0	0.4	0.59	0.86	0.52	0.2	0.32	0.22	0.17
v-10-23-1-16	0.59	0.22	0.44	1.11	0.35	0.2	0.24	0.22	0.15
v-10-23-1-21	1.27	0.69	0.85	0.86	0.91	0.17	0.24	0.2	0.17
v-10-23-1-23	0.81	0.69	0.81	0.98	0.53	0.22	0.41	0.27	0.36
v-10-23-1-24	0.57	0.58	0.58	0.69	0.65	0.06	0.16	0.3	0.28
v-10-28-7-22	0	0.59	1.04	1.38	0.44	0.15	0.21	0.22	0.33
v-10-28-7-22-cpy	0	0.72	0.69	1.33	0.44	0.11	0.21	0.17	0.32
v-10-28-7-23	1.15	0.62	0.75	0.82	0.53	0	0.07	0.06	0.14
v-10-28-7-26	1.17	0.42	0.5	0.83	0.67	0.11	0.31	0.26	0.38
v-10-28-7-29	0.81	0.74	1.01	1.14	1.06	0.29	0.28	0.44	0.39
v-10-28-7-31	0.74	0.69	0.79	1.33	0.94	0.11	0.21	0.27	0.36
v-10-28-7-33	1.13	0.58	1.06	0.42	0.99	0.18	0.35	0.25	0.33
v-10-28-7-36	1.56	0.67	0.42	0.96	0.67	0.17	0.27	0.31	0.31
v-10-29-4-20	0.81	1.19	0.89	1.63	1.26	0.16	0.4	0.19	0.44
v-10-29-4-22	0.81	0.62	1.04	0.62	0.72	0.17	0.22	0.22	0.43

Fig. 9.6 Vibration disagreement scores for the five rating scales and the four facets (part B: vibrations 60–120)

	bumpy	complex	cont.	discont.	dynamic	firm	flat	grainy	irregular	long	pointy	rampdwn	rampup	regular	rough	short	simple	smooth	soft	spiky	springy	ticklish	wavy
agitating	0.16	0.36	0.16	0.46	0.39	0.21	0.04	0.29	0.36	0.2	0	0	0.23	0.3	0.67	0.3	0.08	0.06	0	0.32	0	0.24	0
angry	0.11	0.1	0.14	0.26	0.25	0.21	0	0.16	0.17	0.15	0	0	0.19	0.19	0.5	0.19	0.04	0	0	0.26	0	0.09	0
annoying	0.16	0.24	0.34	0.35	0.42	0.22	0.09	0.39	0.25	0.21	0	0.09	0.37	0.24	0.66	0.31	0.11	0.03	0.03	0.22	0	0.15	0
boring	0.14	0	0.37	0.13	0.14	0.14	0.47	0.36	0.04	0.24	0	0.2	0.09	0.24	0.11	0.19	0.39	0.28	0.27	0.05	0	0	0.1
calm	0.23	0.06	0.22	0.39	0.15	0.08	0.43	0.2	0.09	0.32	0.05	0.27	0.13	0.48	0.05	0.25	0.54	0.62	0.58	0.1	0.04	0	0.05
comfortable	0.5	0.21	0.26	0.52	0.37	0.03	0.22	0.26	0.27	0.21	0.04	0.19	0.16	0.4	0.09	0.26	0.4	0.55	0.56	0.08	0	0	0.07
creepy	0	0.13	0.13	0.02	0.07	0.12	0.12	0.11	0.06	0.07	0	0	0.13	0	0.05	0.07	0.05	0	0	0	0	0.2	0
energetic	0.44	0.1	0	0.29	0.11	0.08	0	0.13	0.04	0.1	0	0	0.1	0.31	0.04	0.2	0.26	0	0.1	0.28	0.13	0	0
familiar	0.29	0	0.19	0.31	0.05	0.1	0.2	0.17	0	0.31	0	0.2	0.08	0.45	0.06	0.26	0.6	0.37	0.48	0.12	0	0	0
fear	0.04	0.16	0.11	0.18	0.21	0.17	0.08	0.23	0.29	0.17	0	0	0.16	0.12	0.27	0.22	0.04	0	0	0.12	0.22	0.35	0.2
funny	0.08	0.3	0.06	0.12	0.19	0	0	0.15	0.21	0	0	0	0.06	0	0.13	0.12	0.04	0.05	0.06	0.14	0.1	0.15	0.1
happy	0.41	0.18	0.05	0.31	0.21	0	0.06	0.2	0.12	0.14	0	0	0.05	0.34	0.1	0.09	0.21	0.2	0.22	0.15	0	0	0.1
interesting	0.48	0.46	0.14	0.45	0.46	0.05	0	0.16	0.33	0.15	0	0.21	0.21	0.23	0.2	0.21	0.23	0.22	0.14	0.23	0.06	0.06	0.05
lively	0.54	0.33	0.03	0.58	0.39	0.12	0	0.16	0.34	0.18	0	0.18	0.4	0.41	0.2	0.24	0.3	0.22	0.18	0.32	0.05	0.04	0
mechanical	0.24	0.32	0.36	0.58	0.51	0.2	0.1	0.44	0.4	0.26	0.03	0.04	0.4	0.34	0.61	0.28	0.28	0.16	0.08	0.24	0	0.08	0.05
natural	0.3	0.09	0.09	0.27	0.13	0.06	0.13	0.12	0.08	0.14	0	0.17	0	0.34	0	0.18	0.34	0.46	0.43	0.1	0	0	0.09
pleasant	0.51	0.18	0.11	0.38	0.18	0	0.14	0.25	0.1	0.15	0	0.05	0.07	0.41	0.03	0.25	0.49	0.44	0.56	0.15	0.06	0	0.12
predictable	0.22	0.03	0.41	0.23	0.22	0.09	0.36	0.37	0.06	0.32	0	0.13	0.24	0.44	0.19	0.14	0.5	0.4	0.37	0.07	0.05	0	0.06
rhythmic	0.55	0.34	0.03	0.54	0.29	0.2	0.04	0.08	0.34	0.25	0.05	0.08	0.03	0.43	0.23	0.12	0.28	0.22	0.21	0.26	0.05	0.04	0
sad	0.04	0.06	0.24	0.08	0.09	0.11	0.5	0.05	0.08	0.31	0	0.4	0.18	0.15	0	0.06	0.26	0.3	0.24	0	0	0	0
strange	0.22	0.39	0.28	0.33	0.43	0.05	0.14	0.28	0.45	0.14	0	0.23	0.28	0.11	0.23	0.25	0.2	0.16	0.1	0.15	0	0.05	0
surprising	0.13	0.48	0.26	0.29	0.47	0.19	0	0.15	0.43	0.04	0	0.18	0.35	0.03	0.33	0.17	0.03	0.11	0.04	0.14	0.09	0.15	0
uncomfortable	0	0.29	0.21	0.27	0.28	0.35	0.06	0.25	0.15	0.21	0	0.17	0.42	0.13	0.45	0.25	0.1	0.04	0	0.14	0	0.21	0
unique	0.28	0.68	0.18	0.41	0.55	0.05	0.05	0.16	0.52	0.18	0.06	0.09	0.14	0.14	0.23	0.14	0.11	0.22	0.11	0.27	0.06	0.11	0
urgent	0.19	0.37	0.29	0.49	0.46	0.25	0.04	0.34	0.35	0.23	0	0.11	0.34	0.31	0.71	0.26	0.17	0.08	0	0.27	0	0.16	0

Fig. 9.7 Co-occurence of sensation$_f$ and emotion$_f$ tags

	bumpy	complex	cont.	discont.	dynamic	firm	flat	grainy	irregular	long	pointy	rampdwn	rampup	regular	rough	short	simple	smooth	soft	spiky	springy	ticklish	wavy
phone	0.18	0.24	0.1	0.28	0.03	0.29	0.07	0.04	0.13	0.2	0	0.14	0.14	0.16	0.39	0.19	0.18	0.08	0	0.26	0	0.09	0
walking	0.05	0.07	0	0.06	0.16	0	0.27	0.06	0.06	0.07	0	0.13	0	0.07	0.09	0.06	0.1	0.06	0.07	0.08	0	0	0
jumping	0.13	0.19	0	0.14	0	0	0	0.05	0.11	0.06	0	0.19	0.19	0.06	0.09	0.06	0.1	0.05	0.06	0.07	0.25	0	0
going away	0.1	0	0	0.02	0.03	0	0	0	0	0.15	0	0.29	0.07	0.07	0	0	0	0.06	0	0	0	0	0
breathing	0	0	0	0.02	0.16	0	0	0.06	0	0.08	0	0.29	0.07	0.06	0	0.06	0.05	0.06	0.07	0	0	0	0
horn	0.04	0.13	0.13	0.1	0.11	0.33	0.11	0.05	0.16	0.13	0	0.11	0.19	0.06	0.17	0.06	0.04	0.05	0	0.07	0	0.17	0
poking	0.22	0.1	0.1	0.23	0.28	0.16	0.08	0.09	0.14	0	0	0.05	0.05	0.17	0.11	0.46	0.26	0.09	0.2	0.23	0	0	0
beep	0.23	0.22	0.19	0.37	0.15	0.2	0.15	0.13	0.24	0.15	0	0.04	0.04	0.28	0.32	0.44	0.26	0.2	0.18	0.2	0	0.12	0.06
pulsing	0.24	0	0	0.18	0.12	0	0.09	0.14	0	0.06	0	0.1	0.06	0.24	0.12	0.17	0.2	0.19	0.39	0	0	0	0.17
snoring	0.12	0	0.28	0.08	0.25	0.09	0.35	0.23	0	0.29	0	0.17	0.28	0.15	0.16	0.06	0.2	0.23	0.16	0	0	0.12	0.15
sliding	0.16	0.18	0.29	0.06	0.09	0	0	0.05	0	0.12	0	0.29	0.29	0.03	0.08	0	0	0.2	0.11	0	0	0.14	0.36
nature	0.08	0	0.12	0.08	0.07	0	0.3	0.15	0.11	0.06	0	0	0.06	0.09	0.04	0.06	0.21	0.15	0.18	0.14	0	0	0
morse code	0.13	0.13	0	0.11	0.13	0.13	0	0	0.06	0.13	0	0	0.1	0.1	0	0	0.05	0.05	0.06	0.08	0.33	0.2	0
riding	0.21	0.06	0.06	0.06	0.11	0	0	0.15	0.11	0.13	0	0.21	0.06	0	0.04	0	0	0	0.06	0	0.25	0.1	0
buzz	0.11	0.1	0.35	0.11	0.27	0	0.22	0.21	0.13	0.31	0	0.15	0.25	0.25	0.19	0.2	0.3	0.17	0.29	0.17	0	0.06	0.07
animal	0.15	0.12	0.24	0.43	0.43	0	0.16	0.28	0.14	0.32	0	0.05	0.35	0.31	0.22	0.2	0.31	0.24	0.27	0.23	0	0.06	0
engine	0.19	0.29	0.49	0.14	0.03	0.06	0.06	0.43	0.22	0.17	0	0.17	0.53	0.12	0.22	0.2	0.16	0.18	0.04	0.09	0	0.07	0
clock	0.13	0	0.06	0.1	0.22	0.24	0.22	0.05	0	0.07	0	0	0	0.13	0	0.13	0.18	0.16	0.19	0.15	0	0	0
musical instrumt	0.14	0.24	0.06	0.26	0.06	0.21	0	0.08	0.21	0.15	0	0.07	0.05	0.09	0.36	0.14	0.04	0.08	0.05	0.26	0	0	0.22
pawing	0.09	0.06	0.06	0.12	0.24	0	0	0	0.05	0.06	0	0	0	0.09	0.04	0.13	0.04	0.21	0.18	0.07	0	0.12	0.07
game	0.32	0.26	0.11	0.42	0.22	0.15	0	0.13	0.34	0.04	0	0.05	0.11	0.26	0.27	0.38	0.21	0.17	0.15	0.12	0	0	0
drums	0.14	0.28	0.05	0.28	0.09	0.21	0	0.08	0.25	0.24	0	0.07	0.12	0.19	0.28	0.14	0.04	0.12	0.09	0.21	0	0.12	0
gun	0.08	0.06	0.12	0.12	0.03	0.11	0	0.3	0.11	0	0.11	0.07	0.12	0.06	0.21	0.3	0.04	0	0.06	0.21	0	0.46	0.07
getting close	0.05	0.07	0.15	0	0.03	0	0.2	0.05	0.05	0.06	0	0	0.07	0	0	0.05	0.05	0	0	0	0	0	0
bell	0.04	0	0.18	0.06	0.4	0	0	0.05	0.05	0.06	0	0.16	0.12	0.09	0.13	0.12	0.26	0.15	0.12	0.07	0	0.09	0
alarm	0.23	0.28	0.19	0.48	0.11	0.2	0.04	0.11	0.35	0.22	0	0.22	0.05	0.29	0.56	0.16	0.18	0.11	0.03	0.23	0	0.09	0.05
heartbeat	0.31	0	0	0.31	0.1	0.14	0	0.04	0	0.29	0	0.2	0.29	0.29	0.04	0.19	0.32	0.24	0.32	0.26	0	0.07	0
SOS	0	0.2	0	0.11	0.38	0.13	0	0	0.11	0.14	0	0.07	0.03	0.03	0.05	0.13	0	0	0	0.23	0	0	0
tapping	0.52	0.17	0.03	0.61	0.3	0.18	0.07	0.21	0.19	0.23	0.04	0.04	0.09	0.57	0.29	0.26	0.38	0.18	0.31	0.36	0.04	0	0.04
coming or going	0.11	0.19	0.29	0.15	0.06	0.14	0.07	0.24	0.09	0.2	0	0.28	0.57	0.14	0.18	0.1	0.07	0.12	0.15	0.05	0	0.18	0
celebration	0.04	0.13	0	0.1	0.03	0	0	0	0.1	0.2	0	0	0.13	0.1	0.13	0.06	0	0.11	0.06	0.07	0	0	0
electric shock	0.04	0	0.18	0.1	0	0.2	0.1	0.29	0.1	0.12	0	0	0.09	0.09	0.29	0.24	0.13	0	0.06	0.07	0	0.29	0

Fig. 9.8 Co-occurence of sensation$_f$ and metaphor$_f$ tags

	bumpy	complex	cont.	discont.	dynamic	firm	flat	grainy	irregular	long	pointy	rampdwn	rampup	regular	rough	short	simple	smooth	soft	spiky	springy	ticklish	wavy
encouragement	0.44	0.42	0.16	0.47	0.45	0.04	0	0.2	0.36	0.1	0	0.08	0.19	0.34	0.26	0.19	0.18	0.29	0.22	0.21	0.05	0.1	0.05
reminder	0.53	0.24	0.2	0.53	0.32	0.11	0.07	0.16	0.22	0.07	0	0.07	0.09	0.46	0.12	0.35	0.41	0.35	0.46	0.28	0.05	0	0.09
congratulations	0.05	0.07	0.07	0	0.03	0	0.14	0.06	0.06	0.07	0	0.14	0	0.03	0	0.05	0.05	0.12	0.07	0	0	0	0
cooldown	0.36	0.17	0.13	0.27	0.21	0	0.12	0.19	0.16	0.22	0.09	0.12	0.17	0.25	0.17	0.13	0.23	0.11	0.21	0.1	0.09	0	0.09
running out of time	0.46	0.28	0.2	0.61	0.52	0.15	0	0.32	0.33	0.18	0	0.09	0.3	0.5	0.49	0.2	0.19	0.25	0.24	0.21	0.04	0.1	0.04
pause	0.13	0.08	0.16	0.3	0.13	0.12	0.23	0.18	0.11	0.13	0.08	0.06	0.08	0.25	0.06	0.42	0.45	0.29	0.44	0.23	0	0	0
milestone	0.54	0.24	0.33	0.46	0.33	0.08	0.08	0.19	0.23	0.06	0.05	0.04	0.09	0.34	0.18	0.48	0.4	0.25	0.24	0.29	0.05	0	0.05
speed up	0.39	0.25	0.14	0.59	0.56	0.1	0	0.25	0.31	0.16	0	0.03	0.41	0.49	0.34	0.16	0.16	0.28	0.16	0.35	0.04	0.15	0
confirmation	0.17	0.05	0.18	0.2	0.11	0.14	0.2	0.08	0.08	0	0	0.07	0.05	0.16	0.07	0.47	0.42	0.24	0.31	0.21	0	0	0.15
halfway	0.2	0.06	0.16	0.12	0.09	0	0	0.09	0.05	0.06	0	0	0.11	0.17	0.04	0.06	0.2	0.28	0.16	0.19	0	0	0.04
above threshold	0.39	0.28	0.08	0.58	0.43	0.17	0.29	0.15	0.26	0.22	0	0.07	0.27	0.44	0.44	0.28	0.21	0.35	0.22	0.26	0.04	0.12	0.04
slow down	0.14	0.1	0.19	0.17	0.17	0	0.23	0.08	0	0.29	0	0.21	0.29	0.27	0.04	0.05	0.18	0.29	0.28	0	0	0	0.06
interval/rep	0.28	0.14	0.17	0.38	0.18	0.14	0.1	0.16	0.13	0.11	0.06	0.14	0.14	0.33	0.31	0.39	0.37	0.19	0.31	0.15	0.06	0.05	0.06
warmup	0.36	0.12	0.15	0.32	0.22	0.05	0.22	0.27	0.18	0.12	0	0	0.23	0.31	0.27	0.19	0.36	0.14	0.19	0.17	0.07	0	0.07
incoming msg	0.31	0.16	0.31	0.26	0.22	0	0.25	0.25	0.23	0.24	0	0	0.08	0.22	0.13	0.36	0.34	0.14	0.27	0.13	0	0	0.07
one minute left	0.46	0.24	0.2	0.5	0.38	0.07	0.04	0.32	0.27	0.09	0	0.11	0.14	0.42	0.32	0.35	0.34	0.21	0.23	0.22	0.05	0.04	0
finish	0.3	0.34	0.3	0.43	0.33	0.24	0.12	0.33	0.23	0.31	0	0.08	0.15	0.35	0.66	0.31	0.3	0.08	0.09	0.16	0	0.09	0
resume	0.29	0.07	0.25	0.34	0.19	0.1	0.03	0.16	0.14	0.04	0	0.05	0.15	0.25	0.15	0.48	0.44	0.26	0.32	0.2	0	0.13	0.06
alarm	0.34	0.33	0.24	0.59	0.54	0.14	0.16	0.31	0.34	0.31	0	0.06	0.38	0.44	0.59	0.26	0.22	0.18	0.14	0.25	0.13	0	0
get ready	0.46	0.29	0.29	0.51	0.36	0.1	0.16	0.24	0.17	0.34	0	0.16	0.21	0.5	0.13	0.29	0.49	0.27	0.36	0.31	0.04	0.07	0.04
start	0.28	0.11	0.11	0.16	0.18	0	0.05	0.05	0.05	0.06	0	0	0.11	0.15	0.04	0.06	0.12	0.1	0.11	0.13	0.18	0	0
battery low	0.23	0.04	0.21	0.35	0.26	0.15	0.19	0.23	0.13	0.11	0	0.19	0.14	0.34	0.12	0.36	0.38	0.32	0.35	0.2	0.03	0.09	0.06
warning	0.5	0.22	0.27	0.6	0.45	0.24	0.08	0.33	0.26	0.16	0	0.08	0.24	0.57	0.31	0.36	0.33	0.27	0.29	0.24	0.03	0.09	0.06
overtime	0.36	0.16	0.24	0.52	0.4	0.2	0.07	0.28	0.28	0.33	0	0.1	0.27	0.45	0.6	0.25	0.28	0.15	0.24	0.2	0	0.08	0.04
below threshold	0.46	0.15	0.12	0.53	0.33	0.11	0.07	0.14	0.14	0.39	0	0.19	0.32	0.5	0.32	0.21	0.3	0.32	0.29	0.19	0.05	0.04	0

Fig. 9.9 Co-occurence of sensation$_f$ and usage$_f$ tags

9.6 Between-Facet Tag Linkages

In this section, we present tag co-occurrence values between the sensation$_f$ facet and emotion$_f$, metaphor$_f$, or usage$_f$ facets (Figs. 9.7, 9.8 and 9.9).

Chapter 10
Supplemental Materials for Chapter 7

Abstract Here, we present additional data on defining emotion controls in Chap. 7. Specifically, we report data that informed our mapping from emotion attributes to sensory attributes (Table 10.1) and then to engineering parameters (Table 10.2). Furthermore, we summarize participant descriptions of our three target emotion attributes (Table 10.3) and provide an example of a base vibration and its derivatives (Figs. 10.1 and 10.2).

© Springer Nature Switzerland AG 2019
H. Seifi, *Personalizing Haptics*, Springer Series on Touch
and Haptic Systems, https://doi.org/10.1007/978-3-030-11379-7_10

10.1   Linkages Between Three Emotion Dimensions and Sensory Attributes of Vibrations

Table 10.1 Three emotion attributes (rows) and their linkages to sensory attributes and tags of vibrations, summarized from Chap. 5. The second column, extracted from a factor analysis in that work, presents the sensory attributes that contribute to the same semantic constructs (a.k.a factors) as the emotion attributes. The last two columns show the most and least correlated tags with each emotion attribute. In Chap. 7, we used six sensory attributes and tags (marked with †): energy, roughness, tempo, discontinuity, irregularity, and dynamism

Emotion attribute	Sensory attribute (factor loading value)	Tags with high correlation (correlation coefficient)	Tags with low correlation (correlation coefficient)
Agitation	Energy (0.9)†	Rough (0.7)	Soft (0.0)
	Roughness (0.8)†	Discontinuous (0.5)	Smooth (0.0)
	Tempo (0.4)†	Dynamic (0.4)†	Flat (0.0)
	Complexity (0.5)	Complex (0.4)	Simple (0.0)
Liveliness	Tempo (0.5)†	Discontinuous (0.6)	Continuous(0.0)
	Continuity (−0.4)†	Bumpy (0.5)	Pointy (0.0)
	Duration (−0.5)	Dynamic (0.4)†	Flat (0.0)
			Ramp down (0.0)
Strangeness	Complexity (0.6)	Irregular (0.5)†	Regular (0.1)
	Continuity (0.3)†	Dynamic (0.4)†	Flat (0.0)
		Complex (0.4)	Simple (0.2)

10.2 Implementation of Sensory Parameters

Table 10.2 Influential sensory attributes from Sect. 7.3.3 (left column), and their implementation with engineering parameters (right column)

Sensory attribute	Implementation
Energy & Roughness	These two sensory parameters are coupled in most vibration actuators including the C2 tactor. Since both of these sensory parameters link to *agitation*, we did not decouple them in our implementation. Past literature links a vibration's energy to its frequency and waveform [1, 2]. We increased frequency and switched to a square waveform
Tempo	We utilized the similarities (in design parameters and hardware) reported for audio and tactile stimuli in the literature [3–6]. Based on the algorithm used for increasing tempo in sound files [7], we shortened duration of pulses and silences without impacting its pitch (frequency)
Discontinuity	Due to lack of prior literature in this area, we qualitatively examined discontinuous vibrations in the *VibViz* library, then modified number and duration of silences in base vibrations, and evaluated them in pilot studies until we converged at the following definition: For discontinuous vibrations, we replaced part of each pulse with silence. For continuous vibrations, we divided the vibration to equal sections and replaced a part of each section with silence
Irregularity	Following a process similar to *discontinuity*, we found linkages between *irregularity* and rhythm variations. Thus, we added silence with a random duration to existing silences in discontinuous vibrations or to random positions in continuous vibrations
Dynamism	Based on *VibViz* vibrations and pilot studies, we defined this as amplitude variation and periodically decreased amplitude of pulses

10.3   Qualitative Description of Emotion Dimensions

Table 10.3 Participant emotion attribute definitions, aggregated for Study 1 and 2. We extracted adjectives and noun phrases and counted participant references to them or their apparent synonyms. The resulting lists are ordered by the most frequent phrases, with the total count presented in parenthesis. Frequently used phrases ($n \geq 4$) for more than one emotion attribute are **bold faced** (and marked with †)

Emotion	Definition pre-interview	Definition post-interview
Agitating	irritating (12), nervous (10), shaking (5), angry (4), uncomfortable (4), unpleasant (3), negative (3), fast (3), random (2), strong (2), constant (2), unbalance (1), provoking (1), attention-getting (1), painful (1), moves up and down (1)	**strong** (25)†, long (6), irritating (5), **fast** (5)†, **non-rhythmic** (5)†, **irregular** (4)†, constant (3), discontinuous (3), aggressive (2), unexpected (2), urgent (2), shaking (2), unpleasant (2), alarming (2), random (2), continuous (2), high frequency (2), frequent pulses (2), different from base (1)
Lively	energetic (11), happy (10), pleasant (7), **strong** (6)†, exciting (5), holidays or party (3), full of life (3), rhythmic (3), upbeat (2), musical (2), alert (2), colorful (1), noisy (1), young (1), confident (1), tickling (1), bright (1), buzzy (1)	**strong** (14)†, **fast** (13)†, rhythmic (10), short pulses (6), discontinuous (3), regular (3), happy (2), upbeat (2), light (1), smooth (1), increase in strength over time (1)
Strange	weird (16), unfamiliar (13), unexpected (6), unpleasant (3), unnatural (3), uncomfortable (2), scary (2), inconsistent (1), disturbing (1), creepy (1), different (1), cautious (1), non-rhythmic (1), patterned vibration (1)	**off-rhythm** (14)†, different from base (8), random pattern (8), unfamiliar (7), **irregular** (6)†, unexpected (4), weird (3), unnatural (2), negative (1), extreme (1), uncomfortable (1), nonsensical (1), long (1), shorter pulses (1), fast (1)

10.4 Specification of Base Vibrations and Their Derivatives

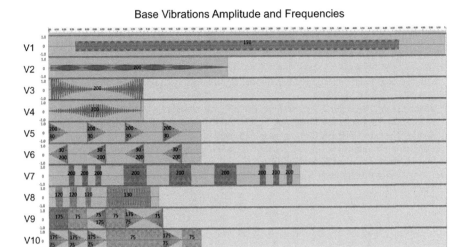

Fig. 10.1 Amplitude and frequency specifications for the 10 base vibrations in our studies

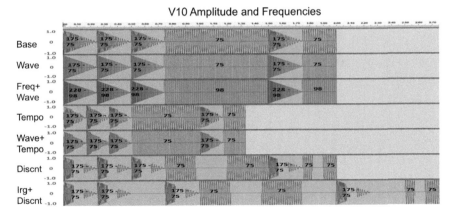

Fig. 10.2 Amplitude and frequency specifications for V10 and its derivatives in Study 1

References

1. O'Sullivan, C., Chang, A.: An activity classification for vibrotactile phenomena, pp. 145–156. Springer, Berlin, Heidelberg (2006). https://doi.org/10.1007/11821731_14
2. Schneider, O.S., MacLean, K.E.: Improvising design with a haptic instrument. In: Proceedings of IEEE Haptics Symposium (HAPTICS '14), pp. 327–332. IEEE (2014)
3. van Erp, J.B., Spapé, M.M.: Distilling the underlying dimensions of tactile melodies. In: Proceedings of Eurohaptics Conference, vol. 2003, pp. 111–120 (2003)
4. Hoggan, E., Brewster, S.: Designing audio and tactile crossmodal icons for mobile devices. In: Proceedings of the 9th ACM International Conference on Multimodal Interfaces (ICMI '07), pp. 162–169. ACM (2007)
5. Brown, L.M., Brewster, S.A., Purchase, H.C.: Tactile crescendos and sforzandos: applying musical techniques to tactile icon design. In: CHI'06 Extended Abstracts on Human factors in Computing Systems (CHI EA '06), pp. 610–615. ACM (2006)
6. Engineering Acoustics, Inc.: C2 tactor https://www.eaiinfo.com/, https://www.eaiinfo.com/tactor-info/. Accessed 21 March 2017
7. SoundTouch: soundtouch algorithm (2016). http://www.surina.net/soundtouch/. Accessed 24 Sept 2016

9783030113780